Laws of Electronic Evidence and Digital Forensics

GAGANDEEP KAUR
Associate Professor
Department of Law, Science and Technology
School of Law
UPES
Dehradun

ANSHIKA DHAWAN
(*Cyber Law Expert*)
Senior Analyst
Ernst & Young LLP
Delhi

PHI Learning Private Limited
Delhi-110092
2025

In fond memory of ***Shri Asoke K. Ghosh*** *(October 1942 – February 2024), Founder Chairman and Managing Director of PHI Learning, whose vision endlessly inspires.*

The Legacy Continues. . . .

Published by Pushpita Ghosh, PHI Learning Private Limited, Rimjhim House, 111, Patparganj Industrial Estate, Delhi-110092 and Printed by Syndicate Binders, A-20, Hosiery Complex, Noida, Phase-II Extension, Noida-201305 (N.C.R. Delhi).

₹550.00

LAWS OF ELECTRONIC EVIDENCE AND DIGITAL FORENSICS
Gagandeep Kaur and Anshika Dhawan

ISBN-978-81-19364-64-0 (Print Book)
ISBN-978-81-19364-89-3 (e-Book)

The export rights of the book are vested solely with the publisher.

Contents

Foreword

We live in a technologically intense world that represents a different dimension of society, where the Internet and Cyberspace serve not only as the most important sources of information but also provide a reliable communication channel. The platform-based ecosystem is the order of today's world. Almost all aspects including crime has a digital dimension or an electronic component in the form of computers and other electronic devices. The borderless, anonymous, and virtual nature of cyberspace and the lack of legal security encourage a setting in which cybercrimes occur. Because of the extensive use of electronic information resources and services, most of the information is either stored or sent in digital form. As a result, these gadgets carry plethora of evidence. What is more thought-provoking is the fragile nature of digital evidence which poses great challenge not only in its collection but preservation too throughout the investigation and trial process.

The Internet and cyberspace are newage phenomena that are evolving every single day. As the technology develops, so are the vulnerabilities, cyber security threats, and thus cybercrimes. Subsequently, as the crimes transition from physical space to virtual space, so do the evidence. The issue stems not only from the nature of the electronic evidence but also from a lack of knowledge, skill, and training in the subject. Electronic evidence and cyberforensics is a new and evolving area that combines both technology and law. The subject, being new, lacks expertise and training, which impedes the justice delivery system. Due to the significant issue is jurisdiction and legal application in the digital environment. Because transnational and complicated nature of crime, appropriate techniques for obtaining digital evidence without infringing on personal integrity rights are provided.

Presently, the subject of electronic evidence and digital forensics is not systemically developed to govern the present scenario. The existing legal framework lacks clarity on the regulation of electronic data. Through a comprehensive examination, the book seeks to contribute to the ongoing discourse on the intersection of technology and law. It aims to offer a nuanced understanding of the challenges posed by the evolving landscape of electronic evidence, while also exploring potential avenues for the establishment of a more robust and adaptive legal framework. By addressing these complexities, the book endeavours to facilitate a more informed and effective approach to handling electronic evidence within the evolving landscape of legal proceedings.

I congratulate Dr. Gagandeep Kaur and Ms. Anshika Dhawan for her textbook on highly relevant subject.

Dr. Gulshan Rai
Former National Cyber Security Coordinator
& Former DG, CERT-In Government of India
in the Office of Prime Minister

Preface

This book is an indispensable resource for professionals and enthusiasts seeking a comprehensive understanding of the evolving landscape of digital forensics and the critical role it plays in the legal and investigative realms. The book adeptly navigates through the complexities of electronic evidence, providing insightful discussions on cutting-edge techniques, methodologies, and tools used in the identification, preservation, and analysis of digital artifacts. Grounded in both theoretical concepts and practical applications, the authors elucidate the challenges posed by digital investigations in the context of cybercrime, online frauds, and other digital offenses. With a keen focus on legal considerations and ethical practices, the book equips readers with the knowledge required to conduct thorough and admissible digital forensic investigations.

I believe that this book on "Electronic Evidence and Digital Forensics" would emerge as a paramount resource, offering a specialized and comprehensive exploration of the intersection between legal precedents and the dynamic field of digital forensics. This book meticulously examines pivotal case laws that have shaped the landscape of electronic evidence, providing readers with invaluable insights into the legal nuances surrounding digital investigations. By dissecting real-world cases, the book not only illuminates the complexities and challenges faced by legal professionals but also underscores the critical role of digital forensics in establishing the admissibility and reliability of electronic evidence in the courtroom. This work serves as an important guide for legal practitioners, forensic experts, and academics seeking to navigate the intricate terrain where technology intersects with the law. Its relevance extends beyond theoretical discussions, offering practical applications and strategic considerations that elevate the proficiency of professionals involved in the evolving realm of electronic evidence and digital forensics.

This book has systematically highlighted landmark decisions by Supreme Courts and High Courts on electronic evidence. Landmark judgments on electronic evidence play a pivotal role in shaping the legal landscape of the digital age, addressing the unique challenges posed by technological advancements. These rulings not only establish legal precedents but also provide essential guidance for courts and practitioners grappling with the intricacies of electronic evidence. By clarifying admissibility criteria, standards for authentication, and the weight given to digital information, landmark judgments contribute to the evolution of a robust and coherent legal framework. They make sure that antiquated procedures do not impede the administration of justice by acknowledging the dynamic nature of technology and its influence on court proceedings. Furthermore, by establishing standards for the gathering, preserving, and presentation of electronic evidence, these rulings ensure a balance between privacy concerns and the necessity of efficient law enforcement. As electronic communication and data storage become increasingly prevalent, these landmark decisions play a crucial role in upholding the integrity of legal processes and fostering confidence in the use of electronic evidence within the judicial system.

This meticulously researched work is a unique blend of two technical subjects namely 'Digital Forensics' and 'Electronic Evidence' that is an essential reference for professionals in law enforcement, legal practitioners, computer scientists, cybersecurity experts, and anyone seeking a deep understanding of the pivotal role played by digital forensics in the pursuit of justice and the protection of digital assets. This textbook is primarily designed for the students of law and forensic science. Besides, it will also be valuable for the lawyers, faculty, judges, and research scholars and professionals in the field of Information Technology, policy, cyber laws, computer science and Internet regulation. I believe that the book will not only prove to be useful resource for all law students but is also a must in all libraries in India and abroad. Any constructive comments for improving the content will be warmly appreciated.

Gagandeep Kaur
Anshika Dhawan

Acknowledgements

I begin by bowing before the Omnipresent who has guided me to take up teaching, research and writing as a career for lifelong learning in Law. My heart could perceive the assiduous, enthusiastic and conscientious efforts imbibed by my co-author Ms. Anshika Dhawan in our book. I wish to quote the following beautiful thoughts by Carlos Ruiz Zafón [The Shadow of the Wind] while writing this manuscript:

"Every book, every volume you see here, has a soul. The soul of the person who wrote it and of those who read it, lived, and dreamed with it. Every time a book changes hands, every time someone runs his eyes down its pages, its spirit grows and strengthens."

Carlos Ruiz Zafón [The Shadow of the Wind]

Billions go to make this great world, but only a few go to make my world beautiful and meaningful. I find no words to acknowledge in so formal manner the love, affection, inspiration and scarifies made by my family. I dedicate all my love and boundless thanks to my little adorable daughters Naira Makhija and Radhya Makhija for their silent inspiration and virtuous support. I wish to manifest my deep-felt gratitude to Mr. Samar Makhija, Mr. Amarnath Makhija, Mrs. Parmod Bala, Mrs. Kuldip Kaur Malhotra, and Sh. V.K. Singh Malhotra (Advocate), as they have enabled me to complete my work smoothly and perfectly.

I would like to express my fathomless appreciation to UPES for providing world-class research facilities. Regards are also due to my learned publisher PHI Learning Pvt. Ltd. who gave due appreciation and recognition to our work.

Gagandeep Kaur

I would like to express my deepest gratitude to my parents, Mr. Satish Dhawan and Mrs. Seema Dhawan, whose unwavering support, encouragement, and endless sacrifices have shaped me into the person I am today.

I am incredibly grateful to my co-author, esteemed professor, and mentor, Dr. Gagandeep Kaur, whose guidance, wisdom, and trust propelled me into co-authoring this book. Dr. Gagandeep Kaur's expertise in the field has been instrumental in shaping the content of this book, and her belief in my potential has been a constant source of inspiration.

I extend heartfelt gratitude to all who shaped my academic journey—my teachers, peers, and friends. Lastly, I extend my sincere appreciation to the readers of this book. I hope that the insights and knowledge shared within these pages will be a valuable resource, contributing to our collective understanding of electronic evidence and digital forensics in India.

Anshika Dhawan

CHAPTER 1

Introduction

Information Technology is one of the fast-growing technologies all over the world.[1] Digital highways are now connecting information islands through advanced communication technology that has revolutionized the modes of transfer and storage of information.[2] Technology is integrated inseparably into our lives to such an extent that our days begin with an alarm clock buzzing on our smartphones and end with surfing through social media or connecting with friends and family on the same device. Digital footprints are accumulated each moment, providing a detailed record of users' online interactions, locations, and preferences.[3]

In the contemporary digital age, technology has transformed every facet of our lives, including the legal landscape. These days, ideas go from the physical to the digital realm through social communication. The usage of Information and Communication Technology (ICT), which encompasses computers, mobile phones, printers, digital cameras, and other gadgets, is central to the virtual world.[4] One of the significant areas profoundly affected is the domain of electronic evidence. The advent of technology has not only transformed the way evidence is gathered and presented

1 Lachman, B.E., Schaefer, A.G., Kalra, N., Hassell, S., Hall, K.C., Curtright, A.E., & Mosher, D.E. (2013). Information Technology Trends. In Key Trends That Will Shape Army Installations of Tomorrow (pp. 171–206). RAND Corporation.

2 Koops, B.J. (2010). Law, Technology, and Shifting Power Relations. *Berkeley Technology Law Journal*, 25(2), 973–1035.

3 Casey, E. (2011). Digital Evidence and Computer Crime (3rd ed.). Elsevier Inc.

4 Mason, S., & Seng, D. (2017). Electronic evidence (4th ed.). Institute of Advanced Legal Studies.

in legal proceedings, however, it has also raised complex jurisprudential challenges.[5] The admissibility and authenticity of electronic evidence[6] can be of critical importance in various contexts, such as legal proceedings, legal agreements, cybercrimes, cybersecurity, or even for e-commerce and online businesses looking to understand consumer behaviour.[7] There are much more opportunities for illegal activities to occur in the virtual world than in the real one, therefore, it is very important to understand the concept of electronic evidence and laws related to it.

According to Black's Law Dictionary, evidence is "something that tends to prove or disprove the existence of an alleged fact."[8] E-evidence is defined as "information recorded electronically on any type of computer device that can be used as evidence in legal action." [9] Electronic evidence may simply be defined as a piece of evidence generated by some mechanical or electronic process.[10] It inculcates but not restricted to e-mails, text documents, spreadsheets, images, graphics, database files, deleted files, data back-ups, located on floppy disks, zip disks, hard drives, tape drives, CD-ROMs, PDAs, cellular phones, microfilms, pen recorders and faxes, etc.[11] Digital or electronic evidence is any type of data kept or transferred on a computer that supports or refutes a theory of occurrence of an offense, or that addresses critical features of the offense like purpose or alibi.[12]

The law relating to evidence, i.e., the Indian Evidence Act was enacted in the year 1872 to bring into the system a codified rule to procure and adduce evidence to decide a dispute in courts. The Evidence Act defines '*evidence*' in Section 3 to mean and include (i) all statements that the Court permits or requires to be made before it by witnesses, in relation to

---

5 Duranti, L., & Stanfield, A. (2021). Authenticating electronic evidence. In S. Mason & D. Seng (Eds.), Electronic Evidence and Electronic Signatures (CMB-Combined volume, 5, pp. 236–278). University of London Press.

6 Leroux, O. (2004). Legal admissibility of electronic evidence. International Review of Law, Computers & Technology, 18(2), 193–220.

7 Schafer, B., & Mason, S. (2017). The characteristics of electronic evidence. In S. Mason & D. Seng (Eds.), Electronic Evidence (4th ed., pp. 18–35). University of London Press. http://www.jstor.org/stable/j.ctv512x65.9.

8 Black's Law Dictionary, 8th Edition, Page 595.

9 Volonino, L. (2003). Electronic evidence and computer forensics. Communications of the Association for Information Systems, 12.

10 Stephen, M., & Daniel, S. (2017). Electronic Evidence. Also see: Goode, S. (2009). The admissibility of electronic evidence. Rev. Litig., 29, 1.

11 Weir, G.R.S., & Mason, S. (2017). The sources of electronic evidence. In S. Mason & D. Seng (Eds.), Electronic Evidence (4th ed., pp. 1–17). University of London Press.

12 Casey, E. (2011). Digital Evidence and Computer Crime (3rd ed.). Elsevier Inc.

matters of fact under inquiry; such statements are called oral evidence; and (ii) all documents produced for the inspection of the Court; such documents are called documentary evidence.[13] Therefore, evidence is broadly divided into two categories, i.e., oral and documentary. The documentary evidence continued to be generated from its conventional source which is the document written in hand or generated by a typewriter or a telegraph. With the advent of 'Computers', 'internet', 'social media', 'chat', 'voice recording', 'emails', and 'social networking data'; documents have largely started being generated through computer printouts. Data being stored on electronic devices became prevalent amongst the general public for storing, creating and communicating data and documents. Evidence Act, thus, amended to make electronic records admissible in evidence.[14]

In the year 2000, Parliament enacted the Information Technology Act 2000 (IT Act) to allow for the admissibility of digital evidence, which amended the Indian Evidence Act 1872 (IEA), the Indian Penal Code, 1860 (IPC) and the Banker's Book Evidence Act 1891. In order to make the electronic evidence admissible, the definition of 'evidence' has been amended to include electronic records. With the introduction of the Information Technology Act 2000, the terms electronic evidence, electronic records, electronic data, etc. were recognized. When the IT Act was enacted, it amended the provisions of legislation like the Indian Penal Code of 1860, the Indian Evidence Act of 1872, the Banker's Book Evidence Act of 1891, and the Reserve Bank of India Act of 1934 to make them more technologically compatible.

Under the Information Technology Act, 2000 (as amended in 2008) the term 'electronic records'[15] means data, record or data generated, image or sound stored, received or sent in an electronic form or microfilm or computer-generated microfiche. The term "Electronic form"[16] means any information generated, sent, received or stored in media, magnetic, optical, computer memory, micro film, computer generated micro fiche or

13 Section 3 of Indian Evidence Act, 1872.

14 Gilliland-Swetland, A. J. (2005). Electronic records management. *Annu. Rev. Inf. Sci. Technol.*, 39(1), 219–253.

15 Section 2(1) (t) of the Information Technology Act, 2000—electronic record means data, record or data generated, image or sound stored, received or sent in an electronic form or micro film or computer generated micro fiche.

16 Section 2(1) (r) of the Information Technology Act, 2000—electronic form with reference to information, means any information generated, sent, received or stored in media, magnetic, optical, computer memory, micro film, computer generated micro fiche or similar device.

similar device. The term "Information"[17] includes data, text, images, sound, voice, codes, computer programmes, software and databases or micro film, or computer generated micro fiche.

The term "All documents prepared for examination by the Court" has been substituted in Section 3 of the Indian Evidence Act by "all documents, *including electronic records* produced available for inspection by the Court." In Section 59, the terms "content of papers" were supplanted by the phrases "Content of electronic documents or records," and Sections 65A and 65B were added to address the admission of electronic evidence.[18]

Legally speaking, any information recorded or transferred in digital form that a person in a court case may utilize at trial is referred to as digital or electronic evidence. Before accepting digital evidence, a court will decide whether it is relevant, authentic, or hearsay and whether a copy is acceptable or the original is required. E-evidence can take the shape of e-mails, digital pictures, ATM transaction logs, word processing documents, instant messaging histories, files saved from accounting applications, spreadsheets, Internet browser history, and databases, among other things.[19]

To modernize the evidentiary practices and enable courts in India to deal with the advances in technology, the Indian Evidence Act, 1872 (hereinafter 'IEA') was amended by virtue of Section 92 of the Information Technology Act, 2000 (hereinafter 'IT Act'). Accordingly, Sections 3 and 59 were amended and Sections 65A and 65B were inserted to incorporate the admissibility of electronic evidence.[20] Any information, be a text, photograph or phone logs, created and stored inside the device is an

17 Section 2(1) (v) of the Information Technology Act, 2000–(v)—information includes [data, message, text,] images, sound, voice, codes, computer programmes, software and data bases or micro film or computer generated micro fiche.

18 Section 59 in The Indian Evidence Act, 1872: Proof of facts by oral evidence.—"All facts, except the [contents of documents or electronic records], may be proved by oral evidence." Jain, N. (2021). Admissibility of E-evidence in India: An overview. *SSRN Electronic Journal*. https://doi.org/10.2139/ssrn.3816724.

19 Santhy, K. V. K., & amp; Irfan, B. M. (2021). Electronic evidence at trial: appreciating digital issues and adjudication process. Natural Volatiles & amp; Essential Oils, 8(5), 2118–2135.

20 Amendment Act 21 of 2000, s. 92 and the Second Schedule: s. 3 "all documents produced for the inspection of the Court" was substituted by "all documents including electronic records produced for the inspection of the Court"; in sec. 59, phrase "content of documents" was substituted by "content of documents or electronic records". Deepa Kharb, "Admissibility of Electronic Records As Secondary Evidence Under Section 65B of The Indian Evidence Act: Recent Judicial Approaches", *Delhi Journal of Contemporary Law*, Vol. I, available at: https://lc2.du.ac.in/DATA/5.pdf, visited on 10 January 2024.

electronic record as per section 2(t) of the IT Act[21] and is a "document" under section 3 of the IEA and therefore admissible before a court as an electronic evidence. Further, according to Section 59, electronic evidence can be proved by documentary evidence only—whether primary or secondary (Section 61) and not by oral evidence.[22] Generally, the computer or digital camera (used for creating or storing electronic data) itself is not produced before the court as evidence, either a printout is taken or CDs/DVDs are prepared to be produced as secondary evidence before the court in place of primary evidence.[23] Thereafter, as per Sections 65A and 65B, the functionality of such computer needs to be established and the person who has transferred these photographs and produced them needs to certify. Further, under Sections 61 to 65, the word "document or content of documents" has not been replaced by the word "electronic documents or content of electronic documents", making the intention of the legislature to be explicitly clear, i.e., not to extend the applicability of Section 61 to 65 to the electronic record.[24]

Digital or electronic evidence is a combination of four elements: (i) computer science; (ii) forensic science; (iii) law; and (iv) behavioural evidence. Computer Science gives the technical knowledge required to comprehend specific features of digital evidence. Forensic Science provides a broad framework for examining any type of digital evidence.[25] Law

21 Section 2(1)(t), the Information Technology Act, 2000: "Electronic record" means data, record or data generated, image or sound stored, received or sent in an electronic form or micro film or computer generated micro fiche. Deepa Kharb, "Admissibility of Electronic Records As Secondary Evidence Under Section 65-B of The Indian Evidence Act: Recent Judicial Approaches", *Delhi Journal of Contemporary Law*, Vol. I, available at: https://lc2.du.ac.in/DATA/5.pdf, visited on 10 January 2024.

22 Under Section 22A, however, it was provided that oral evidence with respect to contents of electronic records is relevant only when genuineness of the record is in question; also Section 45A was inserted in the IEA, which provides that opinion of Examiner of Electronic Evidence will be relevant fact when the Court has to form an opinion on any matter relating to information in electronic form. Also see: Deepa Kharb, "Admissibility of Electronic Records As Secondary Evidence Under Section 65B of The Indian Evidence Act: Recent Judicial Approaches", *Delhi Journal of Contemporary Law*, Vol. I, available at: https://lc2.du.ac.in/DATA/5.pdf, visited on 10 January 2024.

23 Kumar Askand Pandey. "Appreciation of Electronic Evidence: A Critique of Judicial Approach", published in 6 RMLNLUJ (2014) (ISSN 09759549).

24 Deepa Kharb, "Admissibility of Electronic Records As Secondary Evidence Under Section 65B of The Indian Evidence Act: Recent Judicial Approaches", *Delhi Journal of Contemporary Law*, Vol. I, available at: https://lc2.du.ac.in/DATA/5.pdf, visited on 10 January 2024.

25 Smith, J. A. (2020), Electronic Evidence in Legal Proceedings: A Comprehensive Review. *Journal of Cyber Law*, 15(3), 123-145. doi:10.1234/jcl.2020.567890.

involves the study of rules and regulations of the country involving legal investigation of digital or electronic evidence. Behavioral Evidence Analysis is a method of systematizing specialized technical knowledge and general scientific approaches to acquire a better understanding of criminal conduct and motive.[26]

Digital forensics, as a field of study and practice, has emerged as an indispensable component in the realm of information technology and law enforcement.[27] Digital forensics, often interchangeably used with terms like computer forensics or cyber forensics, can be broadly defined as the application of scientific techniques to collect, analyze, and preserve electronic evidence in a manner that is admissible in a court of law. [28] This multifaceted discipline encompasses a range of activities, from the retrieval of data on storage devices to the examination of network traffic and the analysis of digital artifacts.[29] Digital forensics employs a systematic and structured approach to ensure the integrity and admissibility of digital evidence. The process typically involves identification, preservation, collection, examination, analysis, and presentation of electronic evidence. Notable methodologies include disk imaging, file carving, and network forensics, each tailored to address specific types of digital investigations. These methodologies, when applied rigorously, provide investigators with the means to reconstruct events and establish the veracity of digital evidence. The ultimate goal of digital forensics is to uncover facts about digital incidents or crimes, enabling investigators to reconstruct events and attribute actions to specific individuals or entities.[30]

Practically speaking 'Cyber forensics' or 'Digital forensics' or 'Computer forensics' is a branch of forensic science that uses knowledge of the internet and digital technology to create digital or electronic evidence for legal purposes. It is concerned with forensic evidence collected by investigators from cyberspace left during the commission of cybercrime, such as a mobile phone or an iPhone, as well as any computer, laptop, and data. Digital forensics is concerned with all digital equipment, whether it be a digital printer, a digital scanner, a digital photocopy machine, a digital watch, or any other digital object, including, but not limited to, a computer

26 Casey, E. (2011). Digital Evidence and Computer Crime (3rd ed.). Elsevier Inc.

27 Årnes, A. (Ed.). (2017). Digital forensics. John Wiley & Sons.

28 Casey, E. (2011). Digital Evidence and Computer Crime (3rd ed.). Elsevier Inc.

29 Carrier, B., & Spafford, E. (2003). Getting Physical with the Digital Investigation Process. *International Journal of Digital Evidence*, 2(2).

30 Nelson, B., Phillips, A., & Enfinger, F. (2008). Guide to Computer Forensics and Investigations. Cengage Learning.

or a mobile device. Computer forensics is the study of computer systems. Digital forensics is an extension of computer forensics that includes not only computers but also any digital electronic technology, such as mobile phones and printers.[31] Thus, Cyber forensics is the science of gathering, inspecting, analyzing, and reporting electronic evidence, particularly in the context of cybercrime.[32] Computer and network-generated traffic is a rich source of evidence for cyber forensic investigation of potential attacks on sensitive information privacy and integrity.[33] Computer forensics is often divided into two stages: (i) The discovery, recovery, storage, and management of electronic data or documents; (ii) The examination, verification, and presentation of electronic evidence in court or during investigations.[34]

Therefore, it is concluded that electronic evidence and cyber forensics are relatively new additions to the legal jurisprudence. Unlike many older forensic disciplines, which were frequently introduced into the trial process with little or no legal debate and scrutiny, electronic evidence has sparked significant and frequently contentious debate among legal professionals.[35] The discipline of digital forensics is dynamic and constantly changing, having broad consequences for the nexus between law and technology.[36] In the face of cybercrimes and incidents, digital forensics helps investigators unearth the truth by navigating the problematic landscape of digital evidence through a methodical and scientific methodology. Digital forensics are an essential instrument in today's legal environment because of the increasing importance of technology in upholding the integrity of digital investigations and guaranteeing justice.

---

31 Parajuli, R. (2021). Cyber Forensics in Nepal: Critical Study of Laws and Judicial Views. *NJA Law Journal*, 15, 81–106.

32 Hayes, D., & Kyobe, M. (2020). The adoption of automation in Cyber Forensics. 2020 Conference on Information Communications Technology and Society (ICTAS). https://doi.org/10.1109/ictas47918.2020.233977.

33 Chhabra, G. S., Singh, V., Singh, M. (2018). Hadoop-based Analytic Framework for Cyber Forensics. *International Journal of Communication Systems*, 31(15).

34 Volonino, L. (2003). Electronic evidence and computer forensics. Communications of the Association for Information Systems, 12.

35 Mason, S., & Seng, D. (2017). Electronic evidence (4th ed.). Institute of Advanced Legal Studies.

36 Duranti, L., & Stanfield, A. (2021). Authenticating electronic evidence. In S. Mason & D. Seng (Eds.), *Electronic Evidence and Electronic Signatures* (CMB-Combined volume, 5, pp. 236–278). University of London Press.

CHAPTER 2

Fundamentals of Digital Devices
Nature, Types and Formats

A Glimpse into the Chapter

- Meaning of a Computer: Technical & Legal Aspect
- Types of Devices
- Operating Systems
- File Formats
- Storage Formats
- Hash Value

This chapter delves into the fundamental concepts that form the basis of digital discovery and the understanding of the digital world. It explores the technical and legal aspects of computers, different types of devices, operating systems, file formats, storage formats, and the importance of hash values in ensuring the integrity of digital evidence.

2.1 MEANING OF COMPUTER

A computer is an electronic machine that can store, organize, and find information, do processes with numbers and other data, and control other machines.[1] The word computer is derived from the Latin word "computare", meaning: a programmable machine or to calculate. The computer means

1 Campbell-Kelly, M. (2009). Origin of computing. Scientific American, 301(3), 62–69. Also see meaning of computer in the Oxford Dictionary https://www.oxfordlearnersdictionaries.com/definition/english/computer.

a device, which is able to compute.[2] A computer now considered to be a multiple purpose device is used for performing a routine task of adding two numbers as well as a complex task of controlling satellites. It is capable of handling things like numbers, characters, images, audio, video, etc. In computer terminology, these entries of facts are called data.[3] A computer can be defined as an electronic machine that receives data, processes that data and then generates an output. The data is processed based on the instructions given by the user. These instructions, tell the computer how and what is to be processed. The computer also works as a storage device as it stores such instructions. To understand in detail, the Section is divided into two aspects: Technical Perspective and Legal Perspective.

2.1.1 Meaning of Computer from Technical Perspective

A computer is a programmable electronic device that processes, stores, and retrieves data. It can perform various tasks based on a set of instructions provided to it. Computers can be found in a variety of forms, including personal computers, laptops, servers, mainframes, and embedded systems in devices like smartphones and tablets. The fundamental components of a computer include a central processing unit (CPU), memory (RAM and storage), input devices (such as a keyboard and mouse), output devices (like a monitor or printer), and a system that allows these components to communicate with each other.[4] A computer is a combination of both hardware and software components. A traditional computer system consists of a CPU, Memory, Input & Output Devices, and Storage.[5]

Central Processing Unit (CPU)

CPU is the brain of the computer that carries out the actual processing. The working of a CPU is as follows:

(i) Instructions and data are fed to the memory through programs.

---

2 Krull, P. N. (1994). The Origin of Computer Graphics within. IEEE Annals of the History of Computing, 16(3), 41. Also see: Davis, M. (2001). *Engines of Logic: Mathematicians and the Origin of the Computer*. WW Norton & Co., Inc.

3 Malik, Krishna Pal. (2023). Information Technology & Cyber Law. Allhabad Law Agency. pp. 5–8.

4 Randell, B. (Ed.). (2013). *The origins of digital computers: selected papers*. Springer. Also see: Ram, B. (2000). *Computer fundamentals: architecture and organization*. New Age International.

5 Ceruzzi, P. E. (2003). *A history of modern computing*. MIT press. Also see: Ram, B. (2000). *Computer fundamentals: architecture and organization*. New Age International.

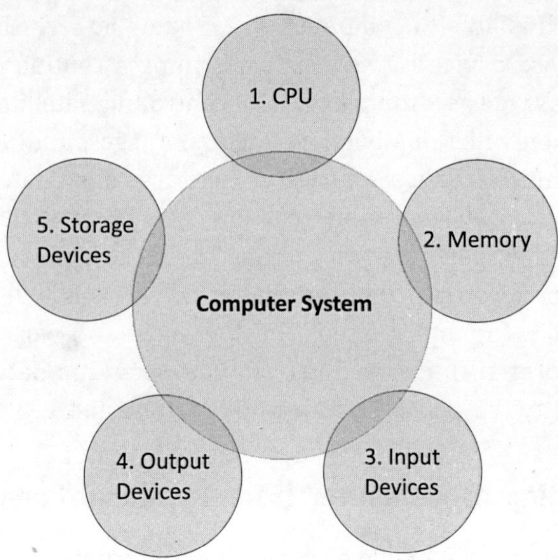

Figure 2.1 Components of a Computer System.

(ii) The data is then fetched from the memory to perform arithmetic and logical operations.

(iii) The result is stored back in the memory as per the given instructions.

The CPU is made up of two major components: (i) the Arithmetic Logic Unit (ALU); and (ii) the Control Unit (CU). All necessary arithmetic and logical operations are carried out by the ALU in accordance with program instructions. The control unit (CU) directs the flow of data through the computer's memory, ALU, and input or output devices. It also supervises the execution of sequential instructions.

Memory

Memory is a component of a computer system that stores data and processing instructions. Here, the term "memory" refers to the main primary memory. To perform read or write operations, the CPU communicates directly with the primary memory. It is further classified into three major types:

(i) **Random Access Memory (RAM):** RAM is volatile, which means that the data stored in it is retained as long as the machine is powered on. The necessary program and data are loaded into RAM for processing every time a computer is turned on or a software program is begun.[6]

6 Memory, R. A., Memory, F., & Memory, S. (2013). RAM.

(ii) **Read Only Memory (ROM):** ROM is non-volatile, its contents are preserved even when the power is switched off. It serves as a compact, quicker permanent storage space for items whose contents are rarely altered.[7]

(iii) **Cache Memory:** A cache is an extremely fast memory that is positioned between the CPU and the primary memory in order to speed up CPU processes. Data copies from commonly used primary memory locations are kept there.

Input Devices

Input devices are objects used to transmit control signals to computers. These devices transform the input data into a digital, i.e., computer-understandable format. E.g.: Mouse, Keyboard, Touch Screen, Voice input, etc.[8]

Output Devices

An output device is a device that receives data from a computer system for display, physical production, etc. It turns digital data into a human-readable format, e.g., Speakers, Printers, Monitor, Projector, etc.[9]

Storage Devices

Storage devices, often known as secondary memory, are non-volatile memory with more storage space than primary memory. Although it is slower and less expensive than the main memory, secondary storage cannot be accessed directly by the CPU; instead, the contents of secondary storage must first be moved into the main memory before the CPU can access them.[10]

2.1.2 Meaning of Computer from Legal Perspective

The Information Technology Act, 2000 defines a computer under Section 2(1)(i) as any electronic, magnetic, optical, or other high-speed data

7 Hartmann, A.K. (2009). *Practical Guide to Computer Simulations (with CD-ROM)*. World Scientific Publishing Company.

8 Sherr, S. (Ed.). (2012). *Input devices* (Vol. 1). Elsevier.

9 Hinckley, K., Jacob, R. J., Ware, C., Wobbrock, J. O., & Wigdor, D. (2014). Input/Output Devices and Interaction Techniques.

10 Schlosser, S.W. (2004). *Using MEMS-based storage devices in Computer Systems* (Doctoral dissertation, PhD thesis, CMU).

processing device or system that performs logical, arithmetic, and memory functions by manipulations of electronic, magnetic, or optical impulses and includes all input, output, processing, storage, computer software or communication facilities which are connected or related to the computer in a computer system or computer network.[11]

According to New Zealand legislation named Search and Surveillance Act 2012[12], the term computer is defined under Section 3(2). According to Section 3(2) of the Search and Surveillance Act 2012, "a computer system encompasses a single computer, interconnected computers, communication links between computers or to remote terminals or other devices, as well as combinations of interconnected computers with communication links. The definition further includes any components of these items, such as input, output, processing, storage, software, communication facilities, and stored data."[13]

From a legal perspective, a computer is often defined as an electronic device that processes and stores data, performing operations according to a set of instructions or programs leading to digital footprints. Legal definitions of computers are typically employed in the context of legislation or regulations addressing computer crimes, data protection, and intellectual property issues. Therefore, a computer is a machine with digital circuits that uses data and performs functions based on instructions leading to the creation of digital pieces of evidence.[14] Computer means an electronic, magnetic, optical, electrochemical, or other high-speed data

11 Section 2(1)(i) Information Technology Act 2000 as amended in 2008: "computer" means any electronic, magnetic, optical or other high-speed data processing device or system which performs logical, arithmetic, and memory functions by manipulations of electronic, magnetic or optical impulses, and includes all input, output, processing, storage, computer software or communication facilities which are connected or related to the computer in a computer system or computer network.

12 New Zealand Legislation—Search and Surveillance Act 2012 (2012 No 24), Retrieved From: https://www.legislation.govt.nz/act/public/2012/0024/latest/DLM2136536.html, visited on 25 January, 2024.

13 Section 3(2) of New Zealand Legislation—Search and Surveillance Act 2012 (2012 No 24): For the purposes of the definition of computer system, a computer is interconnected with another computer if it can be lawfully used to provide access to that other computer— (a) with or without access information; and (b) whether or not either or both computers are currently turned on; and (c) whether or not access is currently occurring.

14 Hayes, P. J., & Berkeley, I. (1997). What is a computer? An electronic discussion. *The Monist*, 80(3), 389–404.

processing device that performs logical, arithmetic, or memory functions by the manipulations of electronic or magnetic impulses and includes all input, output, processing, storage, or communication facilities that are connected or related to the device.[15]

Many legal systems have statutes addressing unauthorized access to computer systems, data interference, and related offences. In these contexts, a computer is often defined broadly to cover a wide range of electronic devices capable of processing data. In legal proceedings, the definition of a computer becomes crucial when dealing with electronic evidence. Courts may define a computer broadly to encompass not only traditional personal computers but also servers, smartphones, tablets, and any other device capable of storing and processing electronic data.

2.2 TYPES OF DEVICES IN COMPUTER SYSTEM: HARDWARE AND SOFTWARE

A computer system is made up of both hardware and software components. All physical components of a computer, such as the keyboard and mouse, are referred to as hardware. Additionally, it contains every component found within the computer, such as the motherboard and circuits, connections, etc. Software is any collection of instructions that directs hardware to perform tasks, e.g., computer programs (Web Browser, Image Viewer, Video Player, etc.).[16]

Each of these hardware or software components holds electronic data that may be potential evidence. The list of devices that constitute a computer or part of a computer system is given in Table 2.1.

---

15 Burks, A. R., & Burks, A. W. (1988). *The first electronic computer: The Atanasoff story*. University of Michigan Press. Also see: Iyer, V., Issadore, D. A., & Aflatouni, F. (2023). The next generation of hybrid microfluidic/integrated circuit chips: recent and upcoming advances in high-speed, high-throughput, and multifunctional lab-on-IC systems. *Lab on a Chip*, 23(11), 2553–2576.

16 Gupta, R. K., Coelho Jr, C. N., & De Micheli, G. (1992, June). Synthesis and simulation of digital systems containing interacting hardware and software components. In *DAC* (Vol. 92, pp. 225–230).

TABLE 2.1 Devices of a Computer System

Sr. No.	*Type of Device*		
1.	Computer in Multiple Shapes & Forms	Desktop	 Image 1: Computer/Desktop
		Laptop	 Image 2: Laptop
2.	Mobile Devices	Smartphone	 Image 3: Smartphone
		Tablet	 Image 4: Tablet
3.	Storage Devices	USB	 Image 5: USB Drive

(*Contd.*)

Sr. No.	*Type of Device*		
		Hard Disks/ Drive	 Image 6: Hard Disk
		External Hard Drive	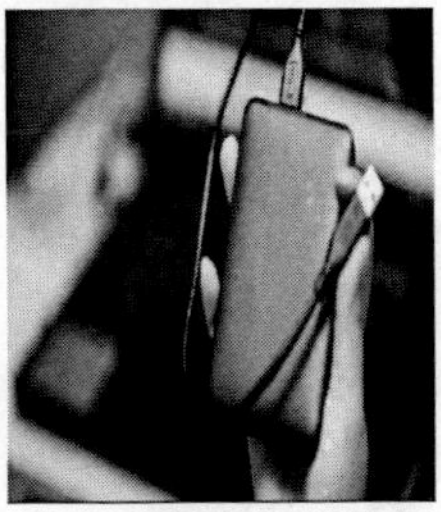 Image 7: External Hard Drive
		CD/DVD	 Image 8: CD
		Memory Card/Chip/ Stick	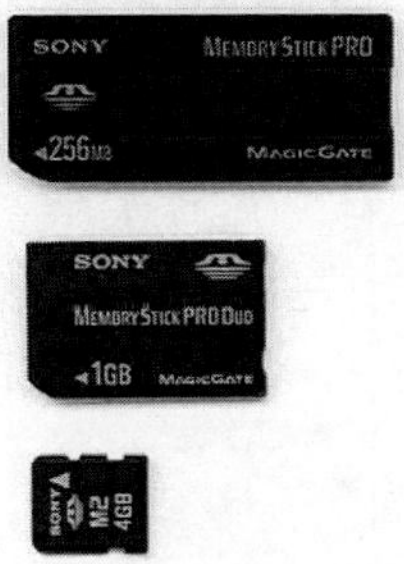 Image 9: Memory Devices
4.	Peripheral Devices	Mouse	 Image 10: Mouse

(*Contd.*)

Sr. No.	Type of Device		
		Keyboard	 Image 11: Keyboard
		Speaker	 Image 12: Speakers
		Microphone	 Image 13: Microphone (Setup)
		Webcam	 Image 14: Webcam
		Media Reader/ Connector/ Hubs	 Image 15: Media Connector Hub

(*Contd.*)

Sr. No.	*Type of Device*		
		Gaming Consoles	 Image 16: Gaming Console
5.	Network Devices	Network Hub	 Image 17: Network Hub
		Network Card	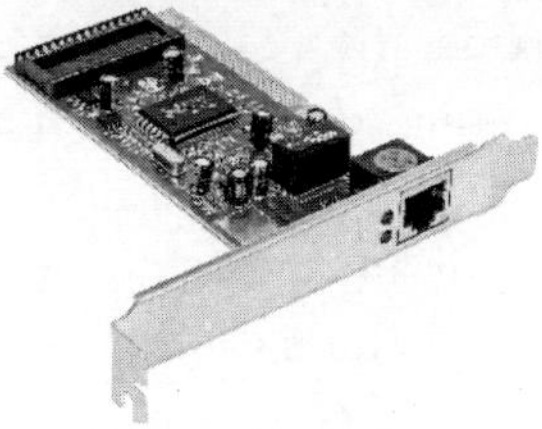 Image 18: Network Card
		Router	 Image 19: Router
		Modem	 Image 20: Modem
		Dongle	 Image 21: Dongle

2.3 OPERATING SYSTEMS

The most crucial piece of software that runs on a computer is the operating system. It controls the memory, operations, software, and hardware of the computer. Additionally, it enables the user to communicate with the computer. The operating system (OS) of a computer oversees all of the software and hardware on the machine. Multiple computer applications are operating at the same time that require access to the computer's central processing unit (CPU), memory, and storage. All of this is orchestrated by the operating system to guarantee that each program receives the resources it requires. Microsoft Windows, macOS, and Linux are the three most popular personal computer operating systems followed by Android by Google, and iOS by Apple are the most commonly used mobile operating systems.

Operating systems serve a wide range of purposes, from personal computing to mobile devices, enterprise-level servers, embedded systems, and specialized tasks like ethical hacking and real-time processing. The choice of an operating system depends on factors such as intended use, hardware compatibility, and specific software requirements.

1. **Microsoft Windows:** Windows is perhaps the most well-known operating system for personal computers. It's celebrated for its user-friendly graphical user interface (GUI) and compatibility with a vast array of software applications. Windows has numerous versions, with Windows 10 and Windows 11 being the most recent. It's widely used for various tasks, from office work to gaming.
2. **MacOS:** Developed by Apple for their Macintosh computers. It's known for its visually appealing and intuitive design, as well as its seamless integration with other Apple products like iPhones and iPads. macOS is popular among creative professionals and those who value a premium computing experience.
3. **Linux:** Linux is an open-source operating system kernel upon which various distributions (distros) are built. Each Linux distro can cater to specific needs and preferences. Linux is widely used for its high level of customization, security features, and stability. It's especially prevalent in server environments and embedded systems.
4. **iOS:** Apple's mobile operating system for iPhones and iPads, is recognized for its user-friendly interface and security features. It is designed for seamless integration with other Apple devices and services.

5. **Android:** Developed by Google, Android is the most popular mobile operating system globally, known for customization and a vast app ecosystem. It's praised for its open nature, allowing extensive customization, and boasts a vast ecosystem of apps on the Google Play Store. Android is used in a wide range of smartphones and tablets.

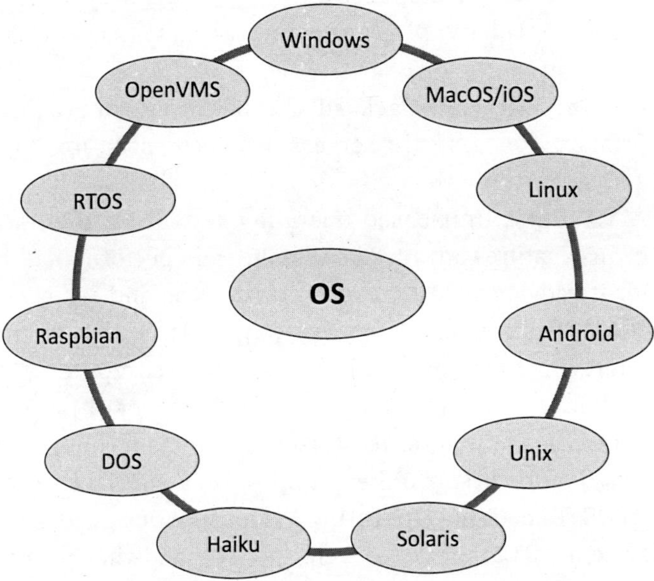

Figure 2.2 Types of Operating System.

6. **Unix:** The precursor to many modern operating systems, Unix is renowned for its multi-user and multitasking capabilities, widely used in server environments. It's the foundation for many modern operating systems and is still prevalent in server environments, particularly in enterprise settings.
7. **Chrome OS:** Designed for lightweight laptops called Chromebooks, Chrome OS mainly runs web-based applications and is ideal for web-focused tasks.
8. **FreeBSD:** An open-source Unix-like operating system is known for advanced networking capabilities, robust security features, and support for various hardware architectures. It's frequently used in server environments.
9. **Solaris:** Developed by Oracle, Solaris is a Unix-based operating system. It's commonly used in enterprise settings, data centers, and other environments that require high levels of reliability and performance.

10. **Haiku:** Inspired by the Be Operating System (BeOS), Haiku aims to provide a fast, efficient, and user-friendly computing experience.
11. **IBM z/OS:** An operating system for IBM mainframe computers, known for its reliability and ability to handle large-scale transactions. It's commonly used in financial institutions and other organizations with high-volume computing needs.
12. **OpenVMS:** Developed by DEC and currently owned by VMS Software, OpenVMS is used in critical systems requiring high reliability.
13. **HP-UX:** Hewlett-Packard's Unix-based operating system, primarily used on HP servers. It's recognized for its scalability and stability.
14. **AIX:** IBM's Unix-based operating system for their server platforms, features advanced scalability and performance. It's known for its advanced scalability, performance, and security features.
15. **DOS (Disk Operating System):** An early command-line operating system used in personal computers, predating Windows. DOS was one of the earliest operating systems for personal computers. It utilized a command-line interface and was the precursor to the graphical user interfaces we use today.
16. **RTOS (Real-Time Operating System):** Designed for embedded systems, RTOS provides deterministic and timely processing of tasks. RTOS is crucial in embedded systems and applications where real-time processing and predictability are paramount. It ensures that tasks are executed with low latency.
17. **Raspbian:** A Debian-based operating system for Raspberry Pi single-board computers. It's tailored to the hardware of these compact devices and is popular among hobbyists and educators.
18. **Kali Linux:** A specialized Linux distribution for penetration testing and ethical hacking, penetration testing, and cybersecurity tasks. It comes preloaded with a suite of security tools for professionals in the field.

2.4 FILE FORMATS

A file format is a standard way that information is encoded for storage in a computer file. It specifies how bits are used to encode information in a digital storage medium. File formats may be either proprietary or free. The file format is the structure of a file that tells a program how to display its contents. For example, a Microsoft Word document saved in the

.DOC file format is best viewed in Microsoft Word. Even if another program can open the file, it may not have all the features needed to display the document correctly. Programs compatible with a file format can give an overview of a file but may not be able to display all the file features. Also, with some programs opening a file format that is not supported may give you garbage.[17] A brief categories of file formats is given in Table 2.2.

TABLE 2.2 File Formats

Category	*Type*	*Extension*	*Description*
Image	Joint Photo-graphic Experts Group	.jpg/ .jpeg	Most common image file format. It compresses the image to minimize storage.
	Graphics Inter-change Format	.gif	Short, animated graphics files.
	Scalable Vector Graphics	.svg	Commonly used for web-site or logo designs.
	Portable Networks Graphic	.png	An image file format that does not change the image quality and size.
	Tagged Image File Format	.tiff/ .tif	It is an image file format that stores high-definition images.
Video	Moving Picture Experts Group Layer Four	.mp4	It is the most used video file format. It may not be able to maintain the video quality of the file.
	Audio Video Interleave	.avi	It was created by Microsoft that maintain video quality.
	Quick Time Movie File	.mov	It was created by Apple that maintain video quality.
	Flash Video Format	.flv	It is a video file format used to stream videos online.
	Advanced Video Coding, High Definition	.avcho	It is a high-quality video file format, generally used by professionals.

(*Contd.*)

17 Computer Hope available at https://www.computerhope.com/jargon/f/file-format.htm visited on 11 January, 2024.

Category	*Type*	*Extension*	*Description*
Audio	MPEG 4 Audio	.mp4	It is a multimedia container format, supporting various audio, video, and subtitle streams for efficient storage and streaming, widely used in digital media applications due to its flexibility and advanced compression techniques.
	MPEG Layer Audio 3	.mp3	It is the most used audio file format that compresses the audio file for easy share and upload.
	Waveform Audio File	.wav	It is a high-quality audio file format, generally used by professionals.
Document	Portable Document Format	.pdf	It is a commonly used document format that does not change the document layout and format.
	Word Document	.doc/.docx	It is a standard document file format in the word processing software.
	Excel Spreadsheet	.xls/.xlsx	The most commonly used file format to store spreadsheets, databases, etc.
	Hypertext Markup Language	.htm/.html	File format used to store data pertaining to web pages.
	Text File	.txt	A simple plain text file.
Presentation	Powerpoint Presentation	.ppt/.pptx	The most used Microsoft PowerPoint software saves presentation files in this file format.
	Open Document Presentation	.odp	It is an OpenOffice file format for presentations.
	Apple Keynote File	.key	File format created by Apple's keynote program to create and share presentations.

2.5 STORAGE FORMATS

Every time one opens up a media file on a computer, there are differing endings in the names of storage formats. Some might be ".mov" while others may be ".avi". These endings represent different storage formats for the file. The same is true for physical data storage as well. These storage formats come in many shapes and forms, such as CDs, external hard drives and USBs. Despite tapes being a very commonly used type of data storage device in past times, they have been outgrown by more recent technologies such as DVDs. Currently, the most commonly used type of data storage device is the USB.[18] Three main generic formats are being extensively used to store data on a computer. Two of these formats are open source, known as Raw and Advanced Forensic Formats (AFF), and the third is Proprietary.[19]

[A] Open Source

[A.1] Raw Format

This is the oldest version of the data format that has been used. It mainly makes a duplicate copy of the disk by performing bit-by-bit copying from one disk to another. For practical purposes to preserve digital evidence, software vendors modified this process with the ability to write bit-stream data to files that create simple sequential files of the media. The output of these files is in raw format.

Raw format outperforms other file formats (like AFF and EWF) in terms of throughput, i.e., has a high transfer data rate between media, yet it has certain drawbacks:

- It is inefficient in using storage capacity; it needs high storage volumes on disk with a minimum capacity equaling the size of the original media.
- It has some efficiency issues when dealing with unhealthy sectors in the media. This means that, when applied on weak media it will have a low level of threshold of retry reads of raw data on these bad media spots.

Many commercial forensic tools have a higher threshold value of retry reads to verify that all relevant data is gathered in a proper way.

18 Data Storage Formats available at: https://clementandjohn.weebly.com/data-storage-and-formats.html visited on 11 January 2024.

19 Carvey, H., & Altheide, C. (2011). *Digital forensics with open source tools.* Elsevier.

Some of the current acquisition tools come with built-in mechanisms to generate raw format acquisitions with validation checksum by using for example Cyclic Redundancy Check (CRC-32), Message Digest 5 (MD5), and Secure Hash Algorithm (SHA-1 or newer versions) and hashing functions.

[A.2] Advanced Forensic Formats (AFF)

This new open-source format has no implementation restrictions and it has started to gain recognition among computer forensics investigators. Although the AFF is open source, many existing software vendors are starting to include it in their tools. AFF format has the following features:

- Using this format, Investigators can create compressed or uncompressed image files:
- It is scalable in storage capacity without any restrictions on file sizes.
- Provides a space in the created image file to store metadata descriptions.
- Has an extendable design with simple concepts.
- It is an open source that runs on multiple computing platforms.
- Has several mechanisms to test internal consistency and self-authentication.

[B] Proprietary Format

Many vendor computer forensics tools can store collected data in specific formats. These formats complement the vendor's analysis software. Vendors' proprietary formats have several advantages compared to other formats such as:

- Efficient in using Disk drive storage space by using options of compressing image files.
- Has built-in mechanisms to split an image into smaller pieces or segments for archiving purposes.
- Has built-in mechanisms for checking data integrity.
- Has metadata features that can be integrated into the image file, such as timestamps.
- They are vendor proprietary formats, which means that it is unable to share an image between different vendor computer forensics analysis tools.

They have some limitations in file size. Typically, forensic tools that use proprietary formats can produce a segmented file having a 2GB maximum segment size.

2.6 HASH VALUE

A hash value, often referred to simply as a hash, is a fixed-size string of characters generated from input data of arbitrary size. Hash functions are mathematical algorithms that take an input (or 'message') and produce a fixed-length string of characters, which is typically a hexadecimal number. The output, or hash value, is unique to the input data, which means that even a small change in the input will result in a significantly different hash value.[20]

To establish the integrity of information a cryptographic hash value, such as MD5 or SHA-1 is calculated so that it can be proven to the courts. Hashing is the process of applying a mathematical procedure to a file/disk storage medium to generate a value that is unique to that file/disk/dataset, and any modifications made to the file/dataset will change/alter the hash value. One of the ways for authenticating any given data set (files, folders, or storage media) that is frequently used in legal proceedings throughout the globe is the hash value. Typically, the hash value is alphanumeric (containing alphabets and numbers). Other hashing algorithms may be used, including SHA256 (secure hash algorithm) and MD5 (Message Digest 5).[21]

The integrity of the media's contents can be demonstrated via a trustworthy hash. A fixed length big integer value (between 80 and 240 bits) reflecting the digital data on the confiscated medium is produced by hashing software. The hash value will change if any alterations are made to the original supporting documentation.[22] Figure 2.3 demonstrates hashing procedure.

Figure 2.3 Hashing Procedure.[23]

---

20 Kumar, K., Sofat, S., Jain, S. K., & Aggarwal, N. (2012). SIGNIFICANCE of hash value generation in digital forensic: A case study. *International Journal of Engineering Research and Development*, 2(5), 64–70.

21 Meaning of MD5, available at: https://www.techtarget.com/searchsecurity/definition/MD5 visited on 11 January 2024.

22 Bellare, M., Goldreich, O., & Goldwasser, S. (1994). Incremental cryptography: The case of hashing and signing. In *Advances in Cryptology—CRYPTO'94: 14th Annual International Cryptology Conference,* Santa Barbara, California, USA August 21–25, 1994 *Proceedings 14* (pp. 216–233). Springer Berlin Heidelberg.

23 Image Retrieved from: https://positiwise.com/blog/what-is-a-hash-function-within-cryptography-quick-guide.

Hash functions are commonly used in computer science and information security for various purposes due to their unique properties and importance:

1. **Data Integrity:** Hashes are used to verify the integrity of data during transmission or storage. By computing the hash value of the original data and comparing it with the hash value of the received data, one can determine whether the data has been tampered with. If the hash values match, the data likely remains unchanged.[24]
2. **Cryptographic Applications:** Hash functions are essential in cryptographic systems for tasks such as creating digital signatures, generating random numbers, and key derivation. They play a crucial role in securing data and ensuring the authenticity and integrity of messages.
3. **Data Verification:** Downloading software from the internet often involves verifying the hash value of the downloaded file against a known hash value provided by the software provider. If the hash values match, it confirms that the download is intact and unaltered.[25]
4. **File Identification:** Many file systems and file-sharing networks use hash values to identify files uniquely. This is particularly useful for checking whether files have been altered or corrupted during transfer.[26]
5. **Digital Forensics:** Hash values play a significant role in digital forensics, helping investigators verify the authenticity and integrity of digital evidence.[27]

Hash values are important in computer science and information security for their ability to provide data integrity, data identification, and security in various applications. They help ensure that data remains

24 Arshad, M. U., Kundu, A., Bertino, E., Ghafoor, A., & Kundu, C. (2017). Efficient and scalable integrity verification of data and query results for graph databases. *IEEE Transactions on Knowledge and Data Engineering,* 30(5), 866–879.

25 Wang, D., Jiang, Y., Song, H., He, F., Gu, M., & Sun, J. (2017). Verification of implementations of cryptographic hash functions. *IEEE Access,* 5, 7816–7825.

26 Gauravaram, P. (2007). *Cryptographic hash functions: cryptanalysis, design and applications* (Doctoral dissertation, Queensland University of Technology).

27 Kumar, K., Sofat, S., Jain, S. K., & Aggarwal, N. (2012). SIGNIFICANCE of hash value generation in digital forensic: A case study. *International Journal of Engineering Research and Development,* 2(5), 64–70.

accurate and secure throughout its lifecycle, and they enable efficient data management and retrieval in a wide range of contexts.[28]

2.7 CONCLUSION

The creation, storage, and accessibility of information has been completely transformed by the widespread use of digital devices, which include computers, smartphones, and Internet of Things (IoT) gadgets. Consequently, these gadgets have evolved into essential resources for law enforcement, attorneys, and forensic specialists gathering digital proof.[29] Electronic evidence generated from these digital devices is becoming a vital part of court cases, criminal investigations, and forensic examinations in the digital age.

Due to their widespread use in contemporary society, smartphones have a plethora of data that may prove invaluable in judicial proceedings. Data extraction and analysis from mobile devices are part of mobile forensics. Using methods like file system analysis, logical extraction, and physical extraction, forensic specialists can obtain call logs, text messages, application data, and location data. Although these techniques yield insightful information, questions regarding user privacy and data security surface, requiring a delicate balance between the demands of investigative work and the rights of individuals.[30]

The increasing prevalence of online activity has increased the use of digital surveillance tools by law enforcement. Electronic evidence collected through monitoring emails, social media interactions, and online communications can be important in criminal investigations. However, potential abuses and violations of data protection rights highlight the importance of legal protection and oversight to prevent undue surveillance.[31]

The Internet of Things (IoT) is creating massive amounts of data that can be used as electronic evidence as they become more and more

28 Aldossary, S., & Allen, W. (2016). Data security, privacy, availability and integrity in cloud computing: issues and current solutions. *International Journal of Advanced Computer Science and Applications*, 7(4).

29 Smith, J. A. (2018). Digital Forensics: Principles and Practices. New York, NY: Academic Press.

30 Brown, M. R., & Jones, S. P. (2020). Mobile Forensics: Advanced Investigative Strategies. Boston, MA: Addison-Wesley.

31 Doe, A. B. (2019). Privacy Concerns in the Age of Digital Surveillance. *Journal of Cybersecurity and Privacy*, 7(2), 134–150. https://doi.org/10.1080/12345678.2019.1234567.

integrated into everyday life. An individual's routines, habits, and even health data can be gleaned from these networked devices, which range from wearable technology to smart home appliances. Ethics concerning ownership of data, consent, and the possible misuse of private information are raised by the collection of such data.[32]

There are new issues for the legal system to deal with when electronic evidence is used in court. Admissibility, authenticity, and dependability of digital evidence are challenges that courts have to deal with. To ensure that people's rights are upheld and law enforcement can successfully combat cybercrime and other illegal activities, legal frameworks must adapt to the increasing complexity of digital investigations.[33]

LET'S RECALL

- A computer is an electronic machine capable of storing, organizing, and processing information, performing numerical and data operations, and controlling other machines. It consists of both hardware and software components, including the CPU, memory, input/output devices, and storage.
- As defined under Section 2(1)(i) of the IT Act 2000, a computer is an electronic, magnetic, optical, electrochemical, or high-speed data processing device that performs logical, arithmetic, or memory functions through electronic or magnetic impulses. This definition encompasses all related input, output, processing, storage, and communication facilities.
- Computers come in various forms, including desktops, laptops, smartphones, tablets, and different storage, peripheral, and network devices.
- An operating system is a crucial piece of software that manages a computer's hardware, software, memory, and operations. Popular operating systems include Microsoft Windows, macOS, Linux, iOS, Android, and others, each tailored for specific devices and use cases.

---

32 Johnson, C. D. (2021). The Internet of Things: Implications for Privacy and Security. *IEEE Transactions on Dependable and Secure Computing*, 18(3), 542–555. https://doi.org/10.1109/TDSC.2021.1234567

33 Legal Foundation for Electronic Evidence. (2022). *Digital Evidence Journal*, 14(1), 45–62. https://www.legalfoundation.org/journals/digital-evidence.

- Files come in various formats, each designed for specific data types. Examples include image formats (e.g., JPEG, GIF), video formats (e.g., MP4, AVI), audio formats (e.g., MP3, WAV), document formats (e.g., PDF, DOCX), and presentation formats (e.g., PPTX, ODP).
- Three primary formats are used to store data: Raw Format, Advanced Forensic Formats (AFF), and Proprietary Formats. Raw Format is an older method for duplicating disk data bit by bit but is inefficient in storage use, AFF is an open-source format with scalability and metadata support, gaining recognition among forensic investigators, and Proprietary formats are used by various forensic tools, providing efficient storage, data integrity checks, metadata features, but are vendor-specific and have limitations.
- Hash functions ensure that small changes in input result in significantly different hash values.
- Hash values are used for data integrity verification, cryptographic applications, data verification, file identification, and digital forensics, as they help maintain the security and authenticity of digital evidence.

Knowledge Check

A. CHOOSE THE CORRECT OPTION

1. According to the Information Technology Act, 2000, how is a computer defined in legal terms?
   (a) Any electronic device with a screen.
   (b) Any device that performs arithmetic functions.
   (c) Any electronic, magnetic, optical, or high-speed data processing device.
   (d) Any device with a keyboard and mouse.
2. What are the three main generic formats used to store data on a computer?
   (a) AFF, EWF, and Raw
   (b) Raw, JPEG, and MP3
   (c) Open, Closed, and Shared
   (d) ASCII, Binary, and Hexadecimal

3. What are the two major components of the CPU?
   (a) Input Unit and Output Unit
   (b) Arithmetic Logic Unit (ALU) and Control Unit (CU)
   (c) Memory and Storage
   (d) RAM and ROM
4. What is the function of a router in a network?
   (a) To display web content
   (b) To connect devices in a local network
   (c) To store data
   (d) To create graphics
5. Which of the following is NOT a commonly used cryptographic hash algorithm mentioned in the text?
   (a) MD5
   (b) SHA–1
   (c) SHA256
   (d) JPEG
6. Which audio file format is commonly used for compressing audio files for easy sharing and uploading?
   (a) WAV
   (b) MP3
   (c) MP4
   (d) AIFF
7. What are peripheral devices in a computer system?
   (a) Devices that store data
   (b) Devices that provide internet access
   (c) Devices used to communicate with the computer
   (d) Devices that are part of the motherboard
8. Which file format is associated with creating and sharing presentations in Microsoft's software suite?
   (a) PDF
   (b) PPT/PPTX
   (c) HTML
   (d) TXT
9. What does the Information Technology Act, 2000 include in its definition of a computer?
   (a) Only the CPU and memory
   (b) All input and output devices
   (c) All processing and storage facilities
   (d) Only the software used in the computer

10. What happens to a hash value if any modifications are made to the original data?
    (a) The hash value remains the same
    (b) The data becomes corrupted
    (c) The hash value changes
    (d) The data is permanently deleted

Answer Key

1. (c) Any electronic, magnetic, optical, or high-speed data processing device.
2. (a) AFF, EWF, and Raw
3. (b) Arithmetic Logic Unit (ALU) and Control Unit (CU).
4. (b) To connect devices in a local network
5. (d) JPEG
6. (b) MP3
7. (c) Devices used to communicate with the computer
8. (b) PPT/PPTX
9. (c) All input, output, processing, storage, computer software, or communication facilities.
10. (c) The hash value changes

B. ANSWER THE FOLLOWING

1. What is a computer?
2. What are the two main components of a computer system? Elaborate.
3. Define a computer according to the Indian Legal System.
4. Why are storage devices important in a computer system, and what are some examples of storage devices?
5. What is a hash value, and how is it used to establish the integrity of information?
6. Describe the three major types of computer memory: Random Access Memory (RAM), Read Only Memory (ROM), and Cache Memory. Explain their functions and characteristics.
7. Name and explain the operating systems commonly used on personal computers.
8. How does the AFF format differ from the Raw format?
9. Explain the role of the Central Processing Unit (CPU) in a computer system and how it carries out processing tasks.

10. In the context of the legal definition of a computer, what are the implications for the inclusion of input, output, processing, storage, computer software, and communication facilities in the definition? How does this broad definition affect the scope of computer-related legal regulations?

CHAPTER 3

Electronic Evidence
Meaning, Nature and Challenges

A Glimpse into the Chapter

- Meaning and Definitions of Electronic Evidence
- Nature of Electronic Evidence
- Sources of Electronic Evidence
- Challenges with Electronic Evidence

In an era where our lives are increasingly intertwined with technology, electronic evidence has emerged as a critical element in legal proceedings, investigations, and the pursuit of justice. From emails and text messages to social media posts and digital documents, our digital footprints can both illuminate and obscure the truth. In a world where the lines between physical and digital realities blur, the need to understand electronic evidence is not just a legal or investigatory imperative; it's a societal one. The stories hidden in data can change lives, alter legal outcomes, and uncover the truth where it might otherwise remain concealed. This chapter is a gateway to the complex and ever-evolving world of digital proof.

3.1 MEANING OF ELECTRONIC EVIDENCE

Electronic evidence, commonly referred to as digital evidence, is a piece of evidence produced by some mechanical or electronic processes that are relevant in proving or disputing a fact or fact in an issue[1]. To put it simply, electronic evidence is evidence that has some electronic or

1 Mueller, C. B., Kirkpatrick, L. C., & Richter, L. L. (2023). *Evidence under the rules: Text, cases, and problems.* Aspen Publishing.

digital significance attached to it that may be used as evidence in a court of law. Further, Electronic or Digital significance can be defined as any data that has been recorded or stored on a computer device, one of its subsets, or a computer network[2]. Electronic Evidence is a combination of two terms: "Electronic" and "Evidence". To understand and define the term Electronic Evidence as a whole, it is pertinent to understand both components individually.

Figure 3.1 Electronic Evidence.

Electronic means something that has or is made up of electronic components that may be microchips, microprocessors, transistors, etc., and operates on electric current, that is electricity. Evidence, on the other hand, is a piece of information that may be used to prove or disprove a fact, proposition, or belief. In the light of The Indian Evidence Act, 1872 "Evidence" is classified into two categories: oral and documentary evidence. The provision states:

"Evidence means and includes:

(i) *all statements that the Court permits or requires to be made before it by witnesses, in relation to matters of fact under inquiry; such statements are called oral evidence;*
(ii) *all documents **including electronic records** produced for the inspection of the Court; such documents are called documentary evidence."*[3]

Initially, the said definition did not include electronic records. The phrases electronic evidence, electronic records, electronic data, and so on became recognized with the enactment of the Information Technology Act of 2000. The amendment led to the replacement of the term *"all documents produced for the inspection of the Court"* with *"all documents **including electronic records** generated for the inspection of the Court"*.

By and large, electronic evidence is any electronically stored information (ESI) that may be used as evidence in a trial or a lawsuit[4], including:

- Documents;
- Emails;

2 Ellen, D., Day, S., & Davies, C. (2018). *Scientific examination of documents: methods and techniques.* CRC Press.
3 Section 3, Indian Evidence Act 1872.
4 Hook, S. A., & Faklaris, C. (2015). Social Media, The Internet and Electronically Stored Information (ESI) Challenges.

- Other files which are stored electronically[5];
- Records that are stored by Internet or network service providers[6].

Although there is no legal definition of *electronic evidence* per se as per Indian Legislation, the explanation of **Section 79A of The Information Technology Act** does shed some light on "electronic form evidence".

It states that for the purposes of this section, "electronic form evidence" means any information of probative value that is either stored or transmitted in electronic form and includes computer evidence, digital audio, digital video, cell phones, digital fax machines.

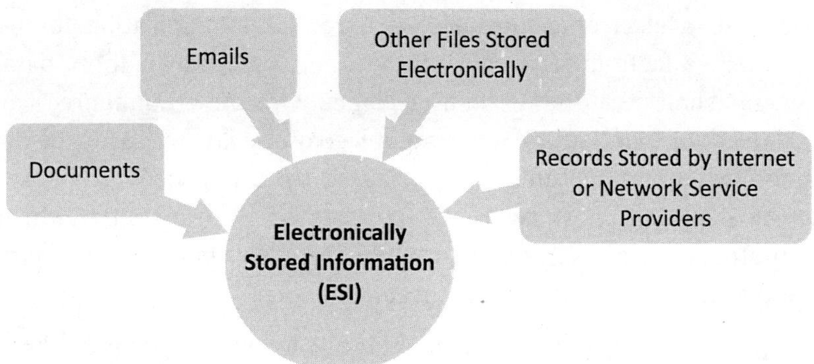

Figure 3.2 Examples of Electronically Stored Information (ESI).

3.2 DEFINITIONS OF ELECTRONIC EVIDENCE

Numerous definitions, interpretations, and viewpoints have been stated on 'electronic' or 'digital evidence' under foreign legislation or offered by authors, academicians, or other researchers. Electronic or Digital evidence is defined as information and data of value to an investigation that is stored on, received, or transmitted by an electronic device'.[7]

The Bulgarian Law on Protection of Competition defines electronic evidence as evidence collected from an undertaking or an association

---

5 Capra, D.J. (2015). Electronically stored information and the ancient documents exception to the hearsay rule: fix it before people find out about it. *Yale JL & Tech.*, 17, 1.

6 Rojas, E.F., & Zeigler, A.D. (2016). *Preserving Electronic Evidence for Trial: A Team Approach to the Litigation Hold, Data Collection, and Evidence Preservation*. Syngress. Also see: https://www.legalmatch.com/law-library/article/electronic-evidence.html#What-Laws-Govern-Electronic-Evidence? Visited on 11 Jan. 2024.

7 RESEARCH ARTICLE Challenges in digital forensics Eva A. Vincze POLICE PRACTICE AND RESEARCH, 2016, VOL. 17, NO. 2,183–194 http://dx.doi.org/10.1080/15614263.2015.1128163.

of undertakings in performing an inspection in electronic form through copying electronic documents and electronic statements.[8]

The UN Convention on Countering the Use of Information and Communications Technologies for Criminal Purposes states electronic evidence means any evidentiary information stored or transmitted in digital form (on an electronic medium).[9] Further, it states that electronic evidence means any data or information generated, stored, transmitted, or otherwise processed in electronic form that may be used to prove or disprove a fact in legal proceedings.[10]

The Cambodian Law on E-Commerce defines electronic evidence as any information, data, or documents that are created, stored, sent, or received in electronic format or electronic communication for being used to prove facts in legal proceedings, and such information, data, or documents shall be authentic in accordance with the e-Commerce Law.

The European Council illustrates electronic evidence as any evidence obtained from data contained in or created by any device, the operation of which depends on software or data stored or transmitted through a computer system or network.[11] Following this, it mentions that the definition of digital evidence has three elements:

(i) First, it is intended to include all forms of evidence that are created, manipulated, or stored in a product that can, in a wide sense, be considered a computer.

(ii) Secondly, it aims to include the various forms of devices by which data can be stored or transmitted, including analog devices that produce an output we presently understand. Ideally, this definition will include any form of device.

---

8 LAW ON PROTECTION OF COMPETITION (Promulgated, State Gazette, Issue 102 of 28.11.2008)
https://www.wipo.int/edocs/lexdocs/laws/en/bg/bg017en.pdf
https://www.lawinsider.com/dictionary/electronic-evidence

9 Article 4(r), United Nations Convention on Countering the Use of Information and Communications Technologies for Criminal Purposes
https://www.unodc.org/documents/Cybercrime/AdHocCommittee/Comments/RF_28_July_2021_-_E.pdf
https://www.lawinsider.com/dictionary/electronic-evidence

10 Article 3(f), UN Convention
https://www.unodc.org/documents/Cybercrime/AdHocCommittee/Second_session/Documents/V22_1AC.pdf

11 Council for Europe
https://search.coe.int/cm/Pages/result_details.aspx?ObjectId=0900001680902e0c
https://ejfs.springeropen.com/articles/10.1186/s41935-021-00234-6#:~:text=As%20defined%20by%20the%20Council,(Council%20of%20Europe%202019).

(iii) Thirdly, the element restricts the data to information that is relevant to the process by which a dispute, whatever the nature of the disagreement, is decided by the adjudicator. This part of the definition includes one aspect of admissibility relevance only—but does not use admissibility' in itself as a deciding criterion, the inadmissibility is well within the authority of the adjudicator.[12]

Egyptian Law interprets electronic evidence as "any electronic information that has a strength or proven value stored, transmitted, extracted, or acquired from computers, information networks, and the like, and that can be collected and analyzed by using special hardware, software, or technological applications"[13]

As notably defined by **Eghon Casey**, Electronic Evidence is any data stored or transmitted using a computer that supports or refutes a theory of how an offense occurred or that addresses critical elements of the offense such as intent or alibi.[14]

Therefore, e-evidence can be broadly defined as any electronically stored information on any type of computing device that can be used as evidence in legal action.[15] It is any probative information stored or transmitted in digital form that may be used at trial. To illustrate, E-Evidence can be found in any digital forms like e-mails, digital photographs, ATM transaction logs, word processing documents, instant message histories, files saved from accounting programs, spreadsheets, internet browser history and databases, etc.[16]

---

12 Council of Europe, Electronic evidence in civil and administrative proceedings, ISBN 978-92-871-8929-5 (JUL 2019). As cited in Shiv, N.N. (2021). Scope of Electronic Evidence in India: Comparative Study (UK, USA & Canada). *Supremo Amicus*, 24, [227]–[234].

13 Egyptian Law No. 175 of 2018.

14 Casey, E. (2011). Digital Evidence and Computer Crime (3rd ed.). Elsevier Inc.

15 Volonino, L. (2003). Electronic evidence and computer forensics. Communications of the Association for Information Systems, 12.

16 Santhy, K. V. K., & amp; Irfan, B. M. (2021). Electronic evidence at trial: appreciating digital issues and adjudication process. Natural Volatiles & amp; Essential Oils, 8(5), 2118–2135.

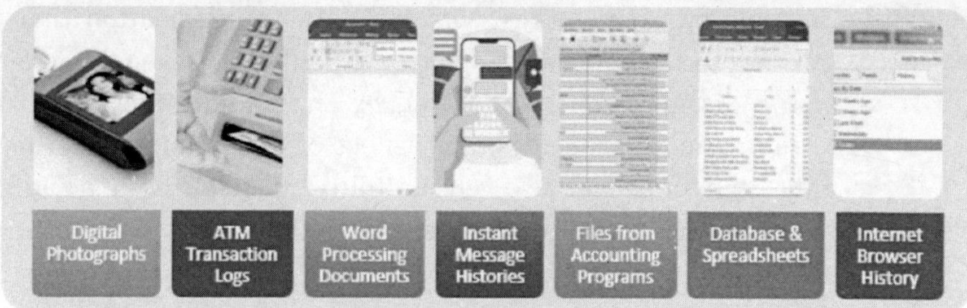

Figure 3.3 Pictorial Representation of Digital Evidence.

3.3 NATURE OF ELECTRONIC EVIDENCE

The nature of electronic evidence refers to the characteristics and attributes of digital information that can be used as proof or documentation in legal proceedings, investigations, or forensic analyses. Any kind of digital artifact created or kept on electronic devices and online platforms is considered electronic evidence. To handle, evaluate, and present electronic evidence in a court of law, legal professionals, forensic specialists, and investigators must have a thorough understanding of the nature of this kind of evidence. Several key aspects contribute to the nature of electronic evidence:

1. **Electronic or Digital Form:** Electronic evidence exists in a digital format, comprising binary data stored on various electronic devices such as computers, smartphones, tablets, servers, and IoT devices. Unlike traditional physical evidence, electronic evidence can be easily copied, manipulated, and transmitted.[17]
2. **Variety or Diversity of Sources:** Electronic evidence is derived from a diverse array of sources, including but not limited to emails, text messages, social media interactions, files stored on devices, internet browsing history, and data generated by IoT devices. Each source presents unique challenges and opportunities for investigation.[18]
3. **Metadata:** Metadata refers to additional information accompanying electronic files, providing details about their creation, modification, and access. This includes timestamps, geolocation

17 Reznik, A. (2019). Metadata: A Digital Investigation Necessity. *Journal of Digital Forensics, Security and Law*, 14(3), 67–80. https://doi.org/10.15394/jdfsl.2019.1594.

18 Casey, E. (2011). Digital Evidence and Computer Crime: Forensic Science, Computers, and the Internet (3rd ed.). Boston, MA: Academic Press.

data, and user identifiers. Metadata is crucial for establishing the authenticity and integrity of electronic evidence.

4. **Volatility and Transience:** Electronic evidence is often volatile and can be easily altered or deleted. The transience of digital data means that timely and secure collection is crucial to preserving its integrity. Digital artifacts may be overwritten or erased if not handled appropriately, emphasizing the need for proper forensic procedures.[19]
5. **Encryption and Security Measures:** Many digital devices and communication platforms use encryption and other security measures to protect sensitive information. The nature of electronic evidence is influenced by these security measures, which can pose challenges in accessing and interpreting the data, especially in cases where decryption is required.[20]
6. **Chain of Custody:** Establishing a secure chain of custody is essential for maintaining the admissibility and reliability of electronic evidence in legal proceedings. Documentation of the handling, storage, and transfer of electronic evidence is critical to ensure its authenticity.
7. **Authenticity and Admissibility Challenges:** Courts often face challenges in determining the admissibility of electronic evidence due to issues such as hearsay, authentication, and the potential for manipulation. Legal professionals must navigate these challenges and adhere to established standards to ensure the credibility of digital evidence.[21]
8. **Global Accessibility:** Electronic evidence can transcend geographical boundaries, as digital information can be stored on servers located in different jurisdictions. This global accessibility introduces complexities related to jurisdictional issues, data protection laws, and international cooperation in legal matters.[22]

---

19 Smith, R. W., & Jones, P. Q. (2020). Challenges in Admitting Electronic Evidence: A Legal Perspective. *International Journal of Law and Information Technology*, 28(2), 145-167. https://doi.org/10.1093/ijlit/eaa017.

20 Taylor, J. M., & Brown, S. A. (2018). The Volatility of Digital Evidence: Challenges and Solutions. Digital Investigation, 25, S34–S42. https://doi.org/10.1016/j.diin.2018.08.006.

21 Greenwald, M. (2017). Encryption and the Law: The Challenge of Balancing Security and Privacy. *Harvard Journal of Law & Technology*, 30(1), 345–372. https://doi.org/10.2139/ssrn.2931705.

22 Greenwald, M. (2017). Encryption and the Law: The Challenge of Balancing Security and Privacy. *Harvard Journal of Law & Technology*, 30(1), 345–372. https://doi.org/10.2139/ssrn.2931705.

In brief, the nature of electronic evidence is characterized by its digital form, diversity of sources, metadata, volatility, encryption challenges, and the need for a secure chain of custody.[23] Legal professionals and forensic experts must navigate these characteristics to effectively leverage electronic evidence while upholding the principles of authenticity, reliability, and privacy. Digital evidence has been described as ephemeral in nature, not reproducible, easily manipulated and inherently suspicious. It is different from evidence created, stored, transferred, and reproduced from a non-digital format.[24]

The nature of e-evidence is latent in the same way that fingerprints and DNA are. It can easily and quickly cross national borders. It is extremely fragile and can be easily altered, damaged, or destroyed. It must be done quickly. E-evidence is a combination of four elements: (i) Computer Science; (ii) Forensic Science; (iii) Law; and (iv) Behavioural Evidence.[25]

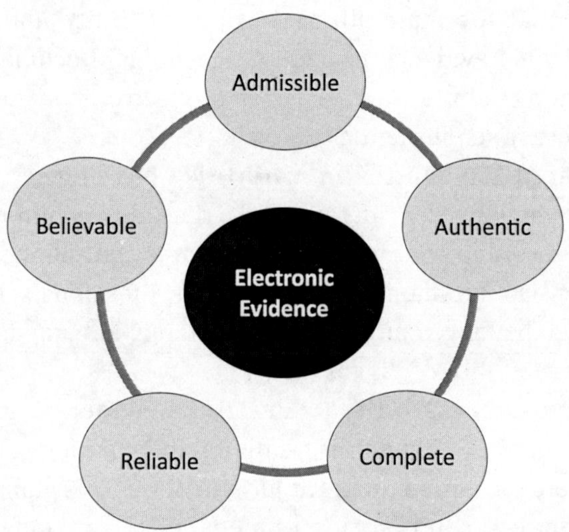

Figure 3.4 Core Features of Electronic Evidence.

Computer Science provides the technical knowledge required to comprehend specific features of digital evidence. The field of forensic

23 Taylor, J. M., & Brown, S. A. (2018). The Volatility of Digital Evidence: Challenges and Solutions. Digital Investigation, 25, S34–S42. https://doi.org/10.1016/j.diin.2018.08.006.

24 RESEARCH ARTICLE Challenges in digital forensics Eva A. Vincze POLICE PRACTICE AND RESEARCH, 2016, VOL. 17, NO.2,183–194 http://dx.doi.org/10.1080/15614263.2015.1128163.

25 Casey, E. (2011). Digital Evidence and Computer Crime (3rd ed.). Elsevier Inc.

science offers a comprehensive framework for analysing all forms of digital evidence. Behavioural Evidence Analysis is a method for systematizing specialized technical knowledge and broad scientific approaches to gain a better understanding of criminal behaviour and motive.[26] A few core features of electronic evidence from the legal perspective are as follows:

(i) **It must be Admissible:** This means that investigators must acceptably present evidence, relate it to the case, and communicate it in a non-biased manner.

(ii) **It must be Authentic:** A digital proof is easy to alter, which raises concerns as to who owns it, or if it is the original, untampered version or not. The authenticity of such evidence should therefore be supported by documents that provide information on its source and relevance to the case. In addition, information such as the source of the evidence, along with the chain of custody must also be presented.

(iii) **It must be Complete:** This means that the electronic evidence should be sufficient to support or refute the fact by linking all relevant points, including the relevant fact, the accused party, and the evidence itself.

(iv) **It must be Reliable:** To demonstrate the reliability of the evidence, forensic experts should extract and handle it while keeping a record of the tasks completed throughout the process. Since the court needs the original proof for future reference, forensic investigations should only be done on copies of the evidence.

(v) **It must be Believable:** The evidence must be presented in a clear and understandable manner.

3.4 SOURCES OF ELECTRONIC EVIDENCE

Today, most of our daily activities are connected to devices or technological products that can generate and store data in digital form. This data could potentially be used as evidence in one way or another.[27] Sources of electronic evidence can be subcategorized into the following:[28]

26 Casey, E. (2011). Digital Evidence and Computer Crime (3rd ed.). Elsevier Inc.

27 Mason, S., & Seng, D. (2017). *Electronic evidence* (p. 422). University of London Press.

28 Kerr, O. S. (2015). An economic understanding of search and seizure law. *U. Pa. L. Rev.*, 164, 591.

Devices

In the past, any type of device with a processing unit was referred to as a "computer." But with the development of modern technologies, gadgets referred to as computers appear to be far away from traditional computers in both shape and function and include digital computation and storage capabilities. These gadgets include gaming consoles, wearable technology (such as fitness trackers and smartwatches), and "smart" home appliances (such as automated central heating systems and smart energy meters).[29] These devices may include:

- Storage Devices like Pen Drives, Hard Disk, CD, etc.
- Handheld Devices like Mobile/Smart Phones, Tablets, etc.
- Desktop that including its hardware, software, and processing units
- Personal Computers or Laptops/Notebooks
- A workstation
- Smart Home Appliances like Smart Refrigerator, Smart AC, Cleaning Robots, Smart Lights, etc.
- Smart Wearables like Smartwatch, Fitness Bands, etc.
- Surveillance Devices like CCTV cameras
- Bluetooth Devices like Headphones, Speakers, Earphones, etc.
- IoT Devices like Alexa, Google Home, etc.
- Other Electronic Goods like Camera, Video Recorders, Smart Alarm Clocks, Smart charging docks, etc.

Network and Internet

Most computers today are permanently connected to a network or other computers, or at least have sporadic connections. Connecting to the Internet, visiting websites, and transferring files between computers, can leave a wealth of electronic evidence. It includes:

- Browsing History
- Information stored on a Router
- Cloud Data
- Servers
- Cellular Networks
- Devices in a shared network, etc.

29 Mason, S., & Seng, D. (2017). Electronic evidence (4th ed.). Institute of Advanced Legal Studies.

Other Sources

Other than the ones stated above, there may be numerous electronic data that are independent of devices and networks, or found on multiple devices or networks, shared, unidentified, but it does exist.[30] Listed below are important sources of Electronic Evidence:

- E-Mails
- Social Media
- Communication over Online Networks like WhatsApp, Telegram, Signal, etc.
- Using E-Commerce websites may leave a trace of the user, especially in cases when an order is placed using the same.
- E-Gaming requires user data to be accessed in the first place.

Pieces of paper with possible passwords, handwritten notes, blank pads of paper with impressions from prior writings, hardware and software manuals, calendars, literature, and text or graphic material printed from the computer that may reveal information relevant to the investigation. These forms of evidence also should be documented and preserved in compliance with departmental policies.

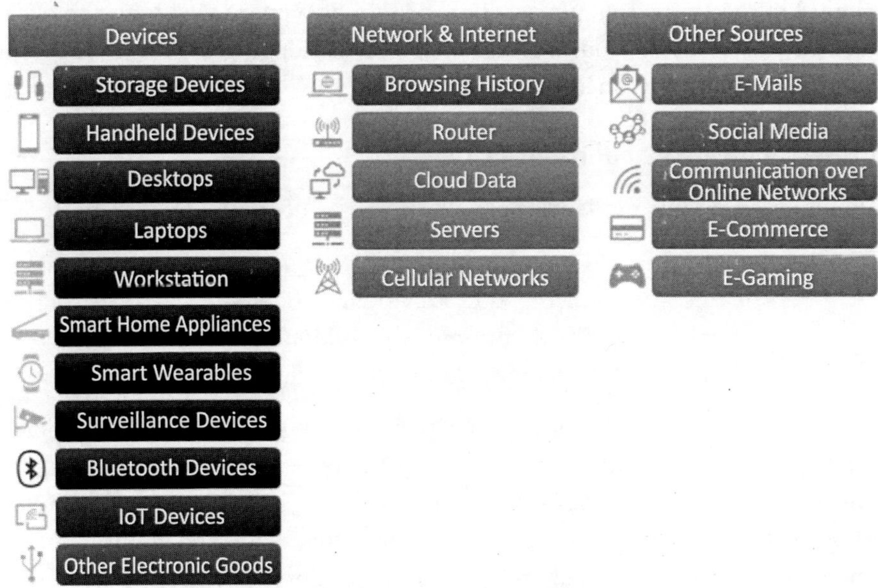

Figure 3.5 Sources of Electronic Evidence.

30 Stephen, M., & Daniel, S. (2017). Electronic Evidence.

3.5 CHALLENGES WITH ELECTRONIC EVIDENCE

Technology is evolving with each passing day and to keep up with it, the law must also evolve. Consequently, electronic evidence, as a synthesis of technology and law, is a work in progress. The subject has significant issues because of its rawness. To name a few, it lacks expertise and training, jurisdiction and legal application in the digital environment are major concerns, cybercrime is transnational and complex, and so on. These issues frequently obstruct the justice delivery system.

Search and Seizure

The manner of collection of digital evidence differs depending on:

- Whether the suspect has already been identified,
- The type of digital evidence,
- Type of case, etc.

Depending on the context, each case requires different procedures, tools, and techniques. When it comes to electronic evidence, investigators need to be aware of what all is included in the ambit of electronic evidence along with where to find that evidence. Additionally, electronic evidence is not limited to what is physically available or what is visible to the eyes but extends to the Virtual world. Thus, it is challenging to identify, search and collect such evidence.

Preservation or Handling of Evidence

Handling electronic evidence normally consists of the following steps:

- Recognition and identification of the evidence
- Documentation of the site of evidence collection
- Collection and preservation of the evidence
- Packaging and transportation of the evidence

It is a challenge because improper handling of electronic evidence might lead to alteration or destruction of evidence because of its fragile nature. Spoliation happens when the person handling the evidence disregards, modifies, or destroys evidence that could be helpful in an ongoing legal dispute. It may be a result of carelessness on the part of the party managing the litigation or from handling and wilful destruction of evidence on the part of the handler.

Presentation and Admissibility

Electronic evidence is extremely important in the investigation process, especially in today's technologically driven environment. It has been a boon to society and the legal system, it does present its own set of challenges. Although electronic evidence is allowed to be used in court proceedings, still it experiences challenges in being proven authentic. Most of the time electronic evidence is challenged in the court due to its integrity. In the absence of proper guidelines and the nonexistence of proper explanation of the collection, and acquisition of electronic evidence gets dismissed in itself. The legal system frequently adopts a lenient stance and does not fully recognize cyber forensics, presenting digital evidence is more challenging than gathering it.

3.6 CONCLUSION

In summary, there are many different sources of electronic evidence, and each one offers particular opportunities and challenges for legal professionals. A deep understanding of these sources is essential for efficient and moral investigative techniques as technology breakthroughs continue to reshape the digital environment. Through careful handling of electronic evidence and adherence to established legal and ethical standards, the legal system can effectively utilize digital artifacts while protecting the rights of parties involved in legal proceedings.[31] The surge in digitalization has transformed the landscape of legal investigations, emphasizing the critical role of electronic evidence. Electronic evidence encompasses a broad array of digital artefacts that play a pivotal role in legal investigations, ranging from criminal cases to civil litigation.[32] Electronic evidence is constantly being kept on distant servers due to the growing dependence on cloud services. Digital artifacts, including images and documents, may be stored on cloud storage platforms like Dropbox and Google Drive. When utilizing and gaining access to cloud-based evidence, attorneys need to take jurisdictional concerns and data protection regulations into account.[33]

31 Miller, A. B., et al. (2019). Internet of Things and Privacy: A Comprehensive Analysis. IEEE Transactions on Dependable and Secure Computing, 17(1), 145–158. https://doi.org/10.1109/TDSC.2019.1234567.

32 Jones, M. R., & Williams, S. P. (2020). Digital Communication and Legal Challenges. *Journal of Digital Investigations*, 15(2), 87–104.

33 Johnson, E. K. (2021). Cloud Storage and Legal Considerations. *Journal of Cyber-security and Privacy*, 19(4), 321–335. https://doi.org/10.1109/TDSC.2021.1234567.

LET'S RECALL

- Electronic evidence is crucial in modern life, impacting legal proceedings and investigations. Electronic evidence is evidence produced by electronic or mechanical processes.
- "Electronic" refers to electronic components like microchips, while "Evidence" is information that proves or disproves a fact. Indian law initially did not include electronic records, but the Information Technology Act of 2000 amended this.
- Digital evidence is ephemeral, not easily reproducible, easily manipulated, and inherently fragile.
- Key Elements of Electronic Evidence include:
    - *Admissibility:* Evidence must be presented in a non-biased manner.
    - *Authenticity:* The origin and source of the evidence must be established.
    - *Completeness:* The evidence should link all relevant points.
    - *Reliability:* Forensic experts should handle the evidence and maintain a chain of custody.
    - *Believability:* The evidence should be presented clearly.
- Sources of Electronic Evidence are:
    - Devices include various electronic gadgets, such as smartphones, tablets, gaming consoles, and smart home appliances.
    - Network and Internet sources encompass browsing history, cloud data, and servers.
    - Other sources include emails, social media, online communication, e-commerce, and gaming.
- Challenges with Electronic Evidence are:
    - Challenges arise due to the evolving nature of technology and law.
    - Key issues include a lack of expertise, jurisdictional complexities, and the transnational nature of cybercrime.
    - Challenges in collecting electronic evidence depend on factors like the suspect's identification and the type of evidence.
    - Preservation is essential but challenging due to the fragility of electronic evidence.
    - Presentation and admissibility are issues, often due to concerns about authenticity and integrity.

A. CHOOSE THE CORRECT OPTION

1. What source of electronic evidence includes information stored on a router?
   (a) E-mails
   (b) Network and Internet
   (c) Social media
   (d) Surveillance devices
2. According to Author Eoghan Casey, electronic evidence is a combination of which elements?
   (a) History, geography, and mathematics
   (b) Computer science, forensic science, law, and behavioral evidence
   (c) Chemistry, physics, and biology
   (d) Literature, music, and art
3. Which legislative act in India recognized the significance of electronic records for legal purposes?
   (a) The Evidence Act
   (b) The Digital Data Protection Act
   (c) The Information Technology Act
   (d) The Cybersecurity Act
4. What is the nature of electronic evidence compared to evidence in non-digital formats?
   (a) Permanent
   (b) Reproducible
   (c) Ephemeral
   (d) Infallible
5. In the context of electronic evidence, what does the term "communication" typically refer to?
   (a) Communication over online networks
   (b) Communication through telepathy
   (c) Communication via carrier pigeons
   (d) Communication via handwritten letters

B. FILL IN THE BLANKS

6. Electronic evidence often faces challenges in being proven authentic and maintaining its _________.

7. Improper handling of electronic evidence might lead to alteration or destruction of evidence, a situation known as _________.
8. In the Indian Evidence Act, the definition of evidence includes both oral and _________ evidence.
9. Electronic evidence must be presented in a clear and understandable manner, making it _________.
10. For electronic evidence to be admissible, it must be _________, reliable, and complete.

Answer Key

1. (b) Network and Internet
2. (b) Computer science, forensic science, law, and behavioral evidence
3. (c) The Information Technology Act
4. (c) Ephemeral
5. (a) Communication over online networks
6. Integrity
7. Spoliation
8. Documentary
9. Believable
10. Authentic

C. ANSWER THE FOLLOWING

1. What are the four elements that make up e-evidence, and why are they important?
2. In your own words, explain what "e-evidence" encompasses.
3. What issues can lead to the dismissal of electronic evidence in a legal case?
4. What is the fundamental concept of electronic evidence, and how is it relevant in legal proceedings?
5. Why is understanding the sources of electronic evidence crucial for investigators and legal professionals?

D. THINK OUTSIDE THE BOX

How can the legal system address the challenges associated with electronic evidence in a rapidly evolving technological landscape?

CHAPTER 4

Understanding Bytes and Beyond Digital Forensics, Types and Techniques

A Glimpse into the Chapter

- Meaning and Definitions of Digital Forensics
- Digital Forensics, Computer Forensics, and Cyber Forensics
- Digital Forensics Tools and Techniques
- Working and Illustration of a few commonly used Tools
- Categorization of Digital Forensics
- Anti-Forensic Techniques

Welcome to the intriguing world of digital forensics, where bits and bytes weave tales of cyber mysteries waiting to be unravelled. This chapter explains the journey to comprehend the intricate field of digital forensics, a subdiscipline of forensic science dedicated to uncovering and investigating the secrets concealed within digital devices. In this digital realm, where every click leaves a trace and every file tells a story, let's understand in unravelling the mysteries encoded in the binary fabric of the cyber world. Let's navigate the landscapes of digital forensics, decoding the language of bytes to understand the secrets they hold. Welcome to Bytes & Beyond, where the digital world becomes a canvas for forensic exploration.

4.1 MEANING AND DEFINITIONS OF DIGITAL FORENSICS

Digital forensic science is a subdiscipline of forensic science that emphasizes the recovery and investigation of information discovered in digital devices. The term digital forensics was coined as a synonym for computer forensics, although its significance and applicability have widened since[1]. In its strictest connotation, digital forensics is the application of computer science and investigative procedures involving the examination of digital evidence—following proper search authority, chain of custody, validation with mathematics, use of validated tools, repeatability, reporting, and possibly expert testimony.[2]

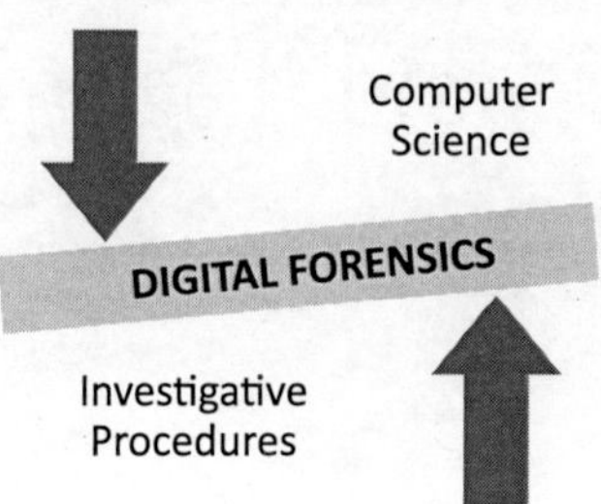

Figure 4.1 Components of Digital Forensics.

Digital forensics per se is a wide subject, it has many definitions, interpretations, and meanings. Generally, it is considered the application of science to the identification, collection, examination, and analysis of data while preserving the integrity of the information and maintaining a strict chain of custody for the data.[3]

Digital Forensics comes from "Digital" & "Forensics".[4] The term "forensic" is derived from the Latin word "forenses" which means "forum." A forum was a public gathering place in early Rome where judicial hearings and debates were held. Forensic science is the general term used for all the scientific processes involved in solving a crime.[5] Forensic science is generally defined as the application of science to the law.[6] Forensics is the practice of gathering, retaining, and analyzing computer-related data

1 Pollitt, M. (2018). Investigating the Cyber Breach: The Digital Forensics Guide for the Network Engineer. Boca Raton, FL: CRC Press. Also see: https://www.eccouncil.org/what-is-digital-forensics/.

2 CNSSI 4009—Committee on National Security Systems (CNSS) Glossary, Release Date: 03/07/2022.
https://www.cnss.gov/CNSS/issuances/Instructions.cfm
https://www.cnss.gov/CNSS/openDoc.cfm?FK4xRINHj3y3SwD3Z3wRvg==

3 https://nvlpubs.nist.gov/nistpubs/Legacy/SP/nistspecialpublication800-86.pdf

4 Årnes, A. (Ed.). (2017). *Digital forensics.* John Wiley & Sons.

5 Kävrestad, J. (2020). *Fundamentals of Digital Forensics.* Springer International Publishing. Also see: https://lawtimesjournal.in/digital-forensics-in-india-an-overview/.

6 https://nvlpubs.nist.gov/nistpubs/Legacy/SP/nistspecialpublication800-86.pdf.

for investigative purposes in a manner that maintains the integrity of the data.[7]

The Scientific Working Group (SWG) defines digital forensics as 'the use of scientifically derived and proven methods towards the preservation, collection, validation, identification, analysis, interpretation, documentation, and presentation of digital evidence derived from digital sources for the purpose of facilitation or furthering the reconstruction of events found to be criminal or helping to anticipate unauthorized actions shown to be disruptive to planned operations.[8]

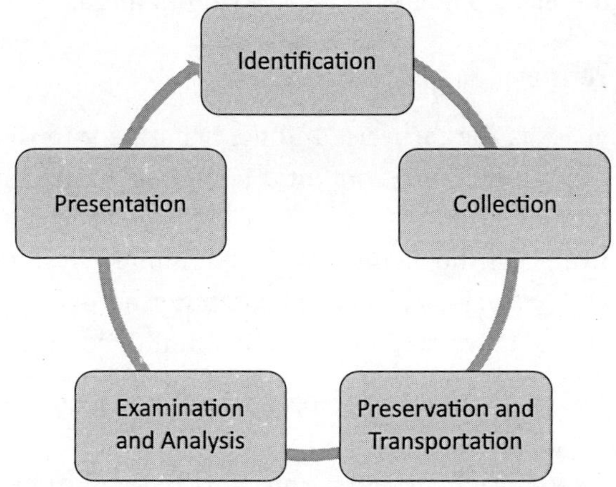

Figure 4.2 Pictorial Presentation of SWG Definition.

4.2 UNDERSTANDING TERMINOLOGY: DIGITAL FORENSICS, COMPUTER FORENSICS, AND CYBER FORENSICS

Although the terms "cyber forensics," "digital forensics," and "computer forensics" are frequently used interchangeably by the public, each term has its own significance and meaning.[9]

Digital Forensics

Digital forensics is a broad field that encompasses the recovery, preservation, analysis, and presentation of digital evidence. It involves the investigation of various types of digital devices and data, such as computers, mobile

7 https://www.cnss.gov/CNSS/openDoc.cfm?FK4xRINHj3y3SwD3Z3wRvg. Also see: Gogolin, G. (Ed.). (2021). *Digital forensics explained.* CRC Press.

8 RESEARCH ARTICLE Challenges in digital forensics Eva A. Vincze POLICE PRACTICE AND RESEARCH, 2016, VOL. 17, NO.2,183–194 http://dx.doi.org/10.1080/15614263.2015.1128163.

9 Parajuli, R. (2021). Cyber Forensics in Nepal: Critical Study of Laws and Judicial Views. *NJA Law Journal,* 15, 81–106.

phones, servers, and storage media. Digital forensics can be applied in both criminal and civil cases, and it includes the examination of data for legal purposes, such as criminal investigations, civil litigation, or corporate investigations. This field aims to recover and analyze digital artefacts and data to answer questions related to how, when, and by whom certain digital activities occurred. Digital forensics is concerned with all types of digital equipment, including but not limited to computers, mobile devices, digital printers, scanners, photocopiers, watches, and other digital objects. It is a division of computer forensics that includes not only computers but also other digital electronic devices like printers and cell phones.

Computer Forensics

Computer forensics is the term used to describe the forensic examination of computer parts and their contents, including hard drives, compact disks, and printers.[10] Computer forensics is a subset of digital forensics, focusing specifically on the examination of computer systems. It involves the collection and analysis of data from computer hardware, software, and related components. Computer forensics is often divided into two stages:

(i) The discovery, recovery, storage, and management of electronic data or documents.

(ii) The examination, verification, and presentation of electronic evidence in court or during investigations.[11]

Computer forensics often deals with issues related to data breaches, computer crimes, unauthorized access, and the investigation of computer-based incidents. The primary goal of computer forensics is to identify and analyze digital evidence related to computer systems and to determine if any malicious or unauthorized activities have taken place.

Cyber Forensics

Cyber forensics is the science of gathering, inspecting, analyzing, and reporting electronic evidence, particularly in the context of cybercrime.[12] It is a field of forensic science that uses expertise in digital and internet technology to produce digital or electronic evidence for judicial use. It is

10 Casey, E. (2011). Digital Evidence and Computer Crime (3rd ed.). Elsevier Inc.

11 Volonino, L. (2003). Electronic evidence and computer forensics. Communications of the Association for Information Systems, 12.

12 Hayes, D., & Kyobe, M. (2020). The adoption of automation in Cyber Forensics. 2020 Conference on Information Communications Technology and Society (ICTAS). https://doi.org/10.1109/ictas47918.2020.233977.

concerned with forensic evidence that investigators have gathered from cyberspace that was left behind during the commission of cybercrime, including any computer, laptop, or data as well as a mobile phone or an iPhone.[13]

Cyber forensics, or cybercrime forensics, is a specialized branch of digital forensics that deals with cybercrimes, which are crimes committed in the digital realm. It focuses on the investigation of criminal activities involving information technology systems, networks, and the Internet.[14] Cyber forensics is concerned with identifying and tracking cybercriminals, understanding their techniques, and preserving digital evidence for legal action. This field is critical in the investigation of cybercrimes like hacking, identity theft, online fraud, and cyberattacks.[15]

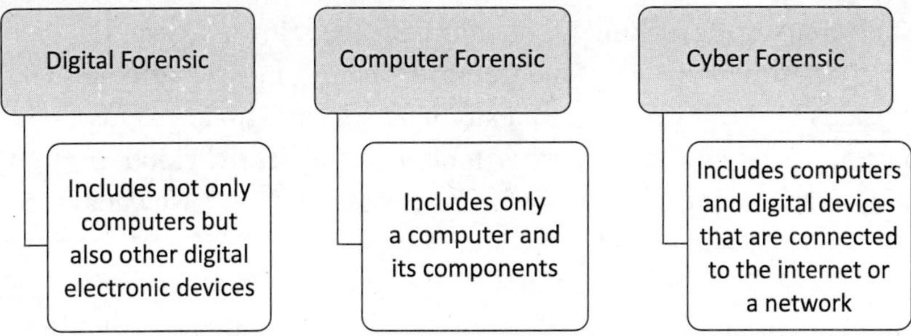

Figure 4.3 Digital Forensic vs. Computer Forensic vs. Cyber Forensic.

4.3 CHALLENGES WITH DIGITAL FORENSICS

Cyber forensics is a new and rapidly developing field that concerns both technology and law. Digital forensics, the practice of uncovering, analyzing, and preserving electronic evidence, plays a critical role in solving cybercrimes and ensuring the integrity of digital investigations. However, this discipline confronts a myriad of challenges that require constant adaptation and innovation. The fact that the subject is new presents significant challenges that may come in the way of the justice delivery system. A notable one is a lack of expertise and training. Due to the complexity and international reach of cybercrime, appropriate methods for gathering digital evidence without infringing an individual's legal rights

13 Harichandran, V. S., Breitinger, F., Baggili, I., & Marrington, A. (2016). A cyber forensics needs analysis survey: Revisiting the domain's needs a decade later. *Computers & Security*, 57, 1–13.

14 Nelson, B., Phillips, A., & Enfinger, F. (2014). Guide to Computer Forensics and Investigations (5th ed.). Boston, MA: Cengage Learning.

15 Reddy, N. (2019). *Practical cyber forensics*. Apress.

are needed.[16] The challenges of digital forensics can be divided into three broad categories:[17]

1. **Technical Challenges:** Different media formats, encryption, steganography, anti-forensics, live acquisition, and analysis are a few examples of technical difficulties.
2. **Legal Challenges:** Jurisdictional problems, privacy problems, and a lack of uniform international law.
3. **Resource Issues:** The amount of data and the length of time needed to gather and analyze forensic media.[18]

Technical Challenges

The rapid evolution of technology presents a significant challenge for digital forensics practitioners. As storage capacities increase, file formats diversify, and new devices emerge, investigators must continually update their skills and tools. The complexity of modern digital systems demands expertise in diverse areas, from mobile forensics to cloud computing, posing a persistent challenge for forensic professionals.[19] Nowadays, nearly every crime has a digital or electronic component in the form of computers or other tools that help to commit the crime. As technology advances, so do security vulnerabilities, cyber threats, and consequently cybercrimes.

- Hardware challenges are related to modulated technology needs and hardware enhancements. According to studies, some criminal suspects may alter the hard disk on their devices before the Cyber Forensic specialist can access it. The globe uses a variety of software, the main challenge here is that every new software comes with new features and abilities that may be complex for a Cyber Forensic expert to understand and gather evidence.
- The amount of data generated by crimes grows along with the frequency of crimes, adding to the difficulty of a digital forensic expert's analysis of such massive data.

---

16 Vincze, E. A. (2016). Challenges in digital forensics. *Police Practice and Research,* 17(2), 183–194.

17 Ghonge, M. M., Pramanik, S., Mangrulkar, R., & Le, D. N. (Eds.). (2022). *Cyber security and digital forensics: Challenges and future trends.* John Wiley & Sons.

18 Al Fahdi, M., Clarke, N. L., & Furnell, S. M. (2013). Challenges to digital forensics: A survey of researchers & practitioners' attitudes and opinions. 2013 Information Security for South Africa. https://doi.org/10.1109/issa.2013.6641058.

19 Nelson, B., Phillips, A., & Enfinger, F. (2014). Guide to Computer Forensics and Investigations (5th ed.). Boston, MA: Cengage Learning.

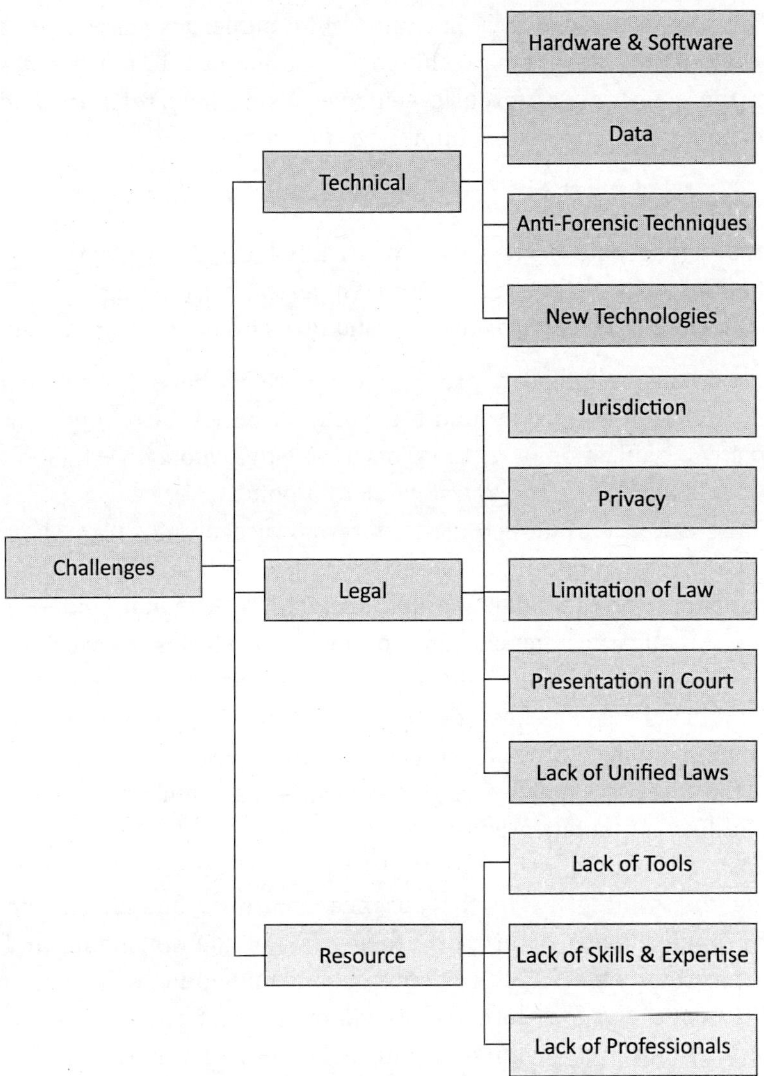

Figure 4.4 Challenges with Digital Forensics.

- Depending on the scenario, the volume of data involved in the case might be large. In that case the investigator has to go through all the collected data in order to gather evidence. It may take more time for the investigation. Since time is a limiting factor, it becomes another major challenge in the field of digital forensics.
- Anti-forensic methods are intended to shield cyberattackers from being identified. The complexity for digital forensic investigators has increased as a result of developments in anti-forensic techniques. There are a number of anti-forensic techniques like data wiping, data hiding, steganography, encryption, etc.

- The computing structure has undergone numerous changes as a result of the current era's technological developments and changes in the collection of forensic evidence. Using new software and technology has created a number of difficulties.

Legal Challenges

The laws governing data protection and privacy have changed in a few different nations around the world. The complexity of gathering forensic evidence is taken into account in some laws and regulations, but many do not.

- The laws governing data protection and privacy have changed in a few different nations around the world. Privacy is also important to any organization or victim. This affects the whole investigation process and limits the investigator to a point.
- There is a lack of unified legislation that incorporates provisions dealing with digital forensics. In India, there are no proper guidelines for the collection and acquisition of digital evidence. The investigating agencies and forensic laboratories are working on the guidelines of their own. Due to this, the potential of digital evidence has been destroyed.[20] Lack of widely recognized standards, which anyone can view and at least get a general idea of the level of expertise of the expert, is the main issue with a field like computer forensics.
- Most of the time electronic evidence is challenged in the court due to its integrity. In the absence of proper guidelines and the nonexistence of proper explanation of the collection, and acquisition of electronic evidence gets dismissed in itself.
- The scope of cyber forensics is wider in itself and the use of various tools and techniques and their different way of working raise lots of issues in front of legal as well as technical experts.
- The widespread use of encryption technologies to secure digital communications poses a formidable challenge to digital forensic investigators. While encryption safeguards user privacy and data integrity, it also obstructs the forensic examination of digital evidence. Balancing the need for investigative access with individual privacy rights remains a contentious issue, requiring a delicate legal and ethical balance.[21]

---

20 https://legaldesire.com/challenges-faced-by-digital-forensics/#_ftnref3.

21 Carrier, B., & Spafford, E. (2003). Getting Physical with the Digital Investigation Process. *International Journal of Digital Evidence*, 2(2), 1–20. https://doi.org/10.2498/i.j.d.e.2003.2.2.

Resource Challenges

- In India, there are only 0.33 forensic experts for every 0.1 million residents, making it difficult to examine crime scenes and produce reports. However, the population ratio of forensic scientists in other countries varies from 20 to 50 scientists per 0.1 million inhabitants, depending on the workload of criminal investigations in various countries.[22]
- Professionals in digital forensics requires adequate education and training because digital technologies are always evolving. A dedication to professional development is necessary to stay up to date with new tools, techniques, and threats. To tackle the ever-changing challenges presented by cybercriminals, it is imperative to have a well-trained workforce.[23]
- With the rapid change in forensic standards, practices, tools and techniques comes a need for ongoing education and training.
- The available tools and equipment may not be able to deal with new and emerging technologies.[24] The lack of tools, procedures, and relevant software or hardware makes it difficult to deal with new technologies, and thus collect digital evidence from the same.

The domain of digital forensics encounters a multitude of obstacles stemming from technological intricacies, encryption, legal implications, global cooperation, and the requirement for ongoing education. It will take a coordinated effort from academics, policymakers, and practitioners to address these issues. The ability of digital forensics to overcome these challenges as technology develops will be crucial to guarantee the success of cybercrime investigations.[25]

4.4 DIGITAL FORENSICS TOOLS AND TECHNIQUES

Digital forensics relies on a variety of specialized tools and techniques to collect, preserve, and analyze digital evidence for investigative and legal

22 https://lawtimesjournal.in/digital-forensics-in-india-an-overview/#_ftn6 *Forensic Science International*; Report, Volume-3, 100215, July 2021.

23 Pollitt, M. (2018). Investigating the Cyber Breach: The Digital Forensics Guide for the Network Engineer. Boca Raton, FL: CRC Press.

24 Karie, N.M., & Venter, H.S. (2015). Taxonomy of challenges for digital forensics. *Journal of forensic sciences*, 60(4), 885–893.

25 Smith, R. W., & Jones, P. Q. (2020). Challenges in Admitting Electronic Evidence: A Legal Perspective. *International Journal of Law and Information Technology*, 28(2), 145–167. https://doi.org/10.1093/ijlit/eaa017.

purposes.[26] These tools and techniques are essential for ensuring the integrity of digital evidence and supporting investigations. These tools can be categorized as follows:

1. **Data Preservation, Duplication, and Verification Tools:** These tools are fundamental for ensuring the integrity and preservation of digital evidence during the forensic process. They are used to create copies (forensic images) of storage devices, such as hard drives, USB drives, and memory cards, without altering the original data. They also ensure data integrity. The key functionalities include[27]:
   (a) *Forensic Imaging:* They create bit-for-bit copies of the entire storage device or specific partitions, including free space and deleted data.
   (b) *Verification:* The tools often calculate cryptographic hashes (MD5, SHA-1, SHA-256, etc.) of the original and duplicate images to verify their integrity.

Examples of Tools under this category:

   (i) *FTK Imager:* This tool is used for disk imaging, and it provides options to create bit-by-bit images.
   (ii) *EnCase:* It offers disk imaging and verification capabilities to maintain data integrity.
   (iii) *dd (command-line utility):* A command-line tool available on Unix-based systems, it can be used for creating disk images and duplicating data.

2. **Data Recovery/Extraction Tools:** Data recovery tools are crucial for retrieving data from digital storage devices, especially when data may have been deleted, lost, or intentionally hidden. These tools help forensic experts recover data by searching for remnants of files in the storage media. Key features include:
   (a) *File Recovery:* They can recover deleted files, even if the file system entries have been removed.
   (b) *Deleted Data Carving:* Some tools use file signatures and heuristics to identify and extract data without relying on file system metadata.

---

26 Carvey, H. (2014). Windows Forensic Analysis Toolkit: Advanced Analysis Techniques for Windows 7 (4th ed.). Burlington, MA: Syngress.

27 Horsman, G. (2019). Tool testing and reliability issues in the field of digital forensics. *Digital Investigation*, 28, 163–175.

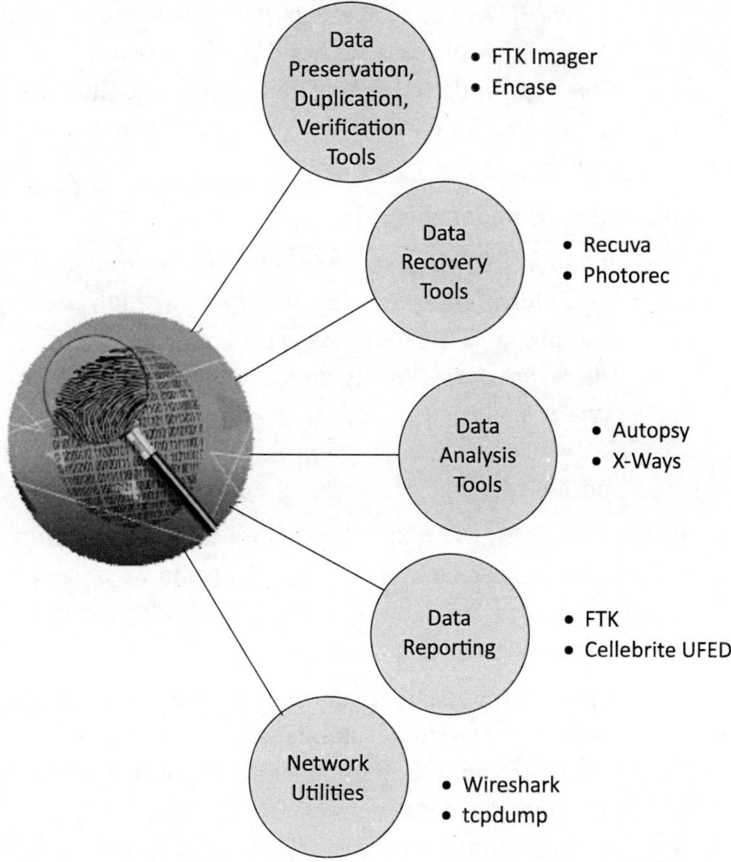

Figure 4.5 Categorization and Examples of Digital Forensic Tools.

Examples of Tools under this category:

(i) *Recuva:* A user-friendly tool for recovering deleted files on Windows systems.

(ii) *PhotoRec:* Primarily used for file recovery and supports multiple file systems and operating systems.

(iii) *GetDataBack:* Effective for recovering data from damaged or formatted drives.

3. **Data Analysis Tools:** Data analysis tools are used to examine the content of digital evidence to uncover insights, patterns, and connections that are relevant to an investigation. These tools often include the following functions:

   (a) *File System Analysis:* They provide an overview of the file system structure and help identify file attributes, timestamps, and file associations.

(b) *Keyword Search:* Investigators can search for specific keywords or phrases within digital evidence.

(c) *Timeline Analysis:* Tools may create a timeline of file access and modification events, aiding in reconstructing events.

Examples of Tools under this category:

(i) *Autopsy:* An open-source digital forensic tool that provides a graphical interface for file system analysis, keyword searches, and timeline analysis.

(ii) *The Sleuth Kit:* Works alongside Autopsy and offers file system analysis capabilities.

(iii) *X-Ways Forensics:* A comprehensive tool for data analysis and file carving.

4. **Data Reporting Tools:** Reporting tools are essential for documenting forensic findings in a clear and organized manner, as these reports are often presented as evidence in court. The key aspects of these tools include:

(a) *Report Templates:* They offer pre-defined templates for creating professional, standardized reports.

(b) *Customization:* Users can tailor reports to meet specific case requirements.

(c) *Incorporating Evidence:* They enable the inclusion of images, logs, and analysis summaries in the reports.

Examples of Tools under this category:

(i) *EnCase:* In addition to imaging and analysis, EnCase provides robust reporting features for documenting forensic findings.

(ii) *FTK (Forensic Toolkit):* Offers reporting capabilities that can be customized to meet specific reporting requirements.

(iii) *Cellebrite UFED (Universal Forensic Extraction Device):* Provides reporting functionality for mobile device forensics.

5. **Network Utilities:** Network utilities are used in network forensics to monitor, capture, and analyze network traffic. These tools assist in identifying security incidents and tracking potential threats. Key functionalities include:

(a) *Packet Capture:* Network analyzers capture network packets, including their contents, headers, and metadata.

(b) *Protocol Analysis:* They dissect and interpret network protocols and application-layer data.

(c) *Network Traffic Visualization:* Some tools provide visual representations of network traffic flow.

Examples of Tools under this category:

(i) *Wireshark:* An open-source network protocol analyzer for capturing and analyzing network packets.

(ii) *tcpdump:* A command-line packet analyzer available on Unix-based systems, often used for network traffic capture.

(iii) *NetworkMiner:* A tool for network packet analysis, which can extract files and information from captured network traffic.

4.5 EXPLAINING THE WORKING OF POPULAR TOOLS WITH ILLUSTRATION

(a) **Autopsy**

- *Purpose:* Primarily designed for hard drive investigations; also used for recovering photographs from camera memory cards.
- *Capabilities:* Identifying relevant data portions, flagging contact information, emails, and files, creating a unified data repository.
- *Users:* Utilized by law enforcement, military, and corporate examiners.

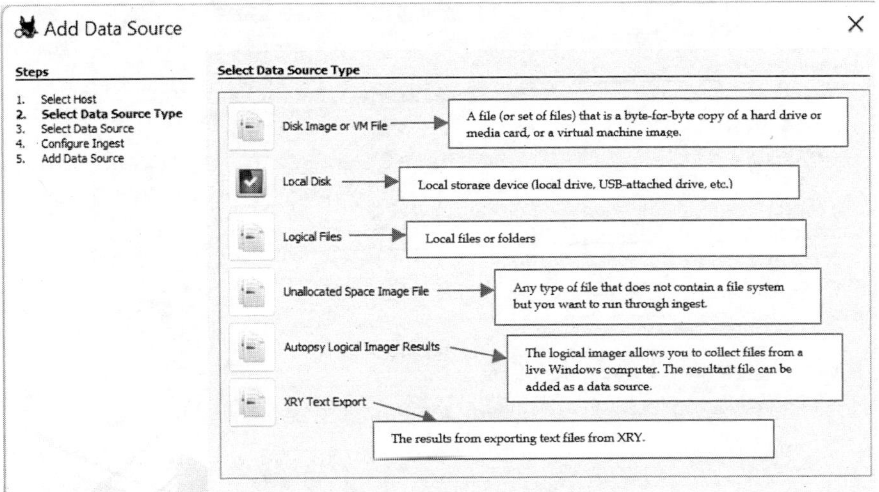

Figure 4.6 Autopsy Tool Interface.

- *Compatibility:* Works on Linux, Unix, macOS, and Windows.
- *Extensibility:* Supports additional functionality through user-developed plugins.
- *Accessibility:* Allows multiple users and provides a centralized approach for uniform access.
- *Ease of Use:* Offers shortcuts and historical tools for efficient processes.

(b) **X-Ways Forensics**

Figure 4.7 X-Ways Logo.

- *Features:* Powerful binary editor providing access to various computer components.
- *Search Tools:* Lightning-fast simultaneous search tools for scanning media, including deleted files.
- *Cloning/Imaging:* Capable of cloning/imaging even undetectable volumes and verifying image file integrity.
- *Media Formats:* Supports various media formats for sector-by-sector copies.
- *Logging:* Generates log files recording damaged sectors during cloning.

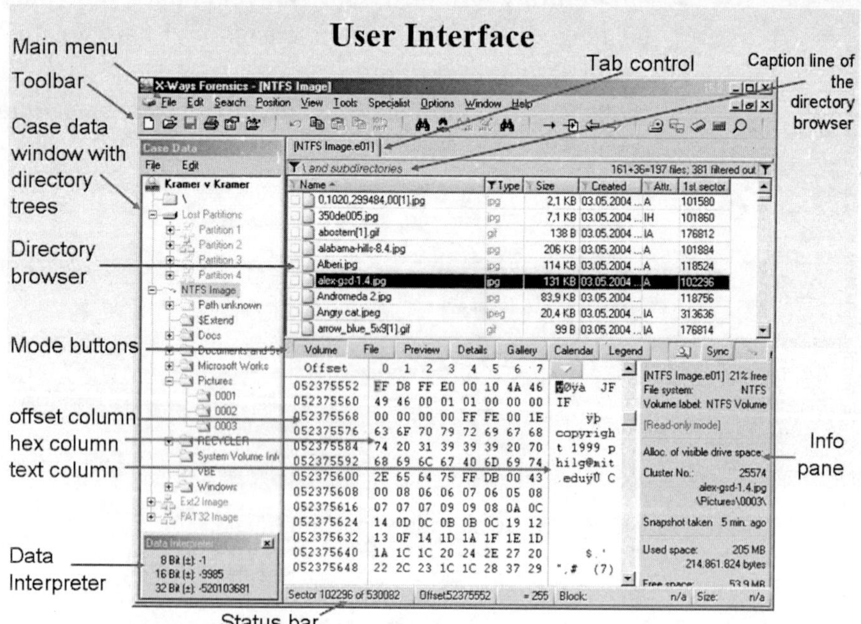

Figure 4.8 X-Ways Tool Interface.[28]

28 Image Retrieved from: https://www.x-ways.net/forensics/.

(c) **Cellebrite UFED**

- *Focus:* Mobile forensics tool for bypassing locks, performing extractions, unlocking devices, and recovering data.
- *Extraction Methods:* Logical, file system, physical extractions for deep data recovery.
- *Device Compatibility:* Extracts data from mobile phones, drones, SIM cards, SD cards, GPS trackers.
- *Methods of Recovery:* Utilizes exclusive bootloaders, automatic EDL capability, Smart ADB, and more.
- *Platform Availability:* Accessible on multiple platforms.

Figure 4.9 Cellebrite UFED Logo.

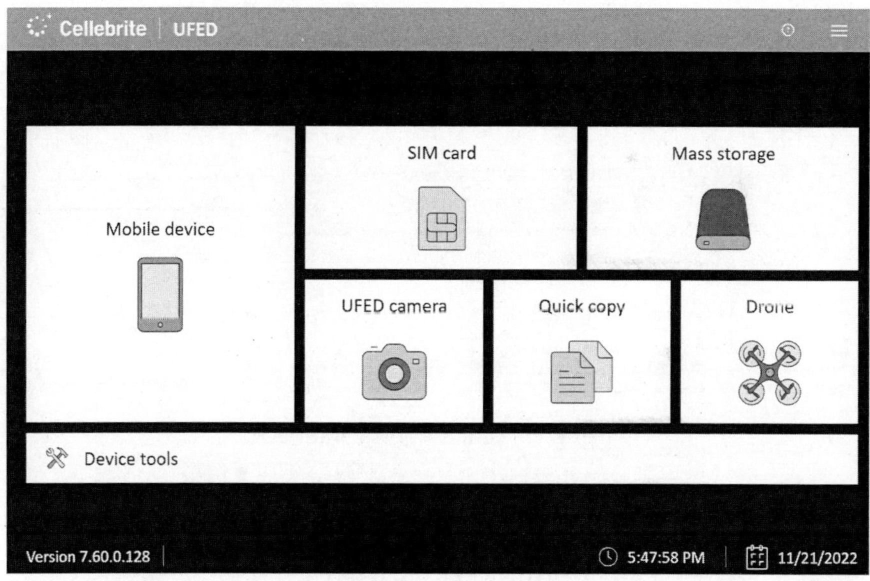

Figure 4.10 Cellebrite Tool Interface.[29]

(d) **EnCase Forensic Investigation Software**

- *Case Management:* Developed by Guidance Software, assists in data and evidence collection.

29 Image Retrieved from: https://cellebrite.com/en/ufed/.

- *Data Collection:* Acquires forensic data from various operating and file systems, including mobile devices.
- *Decryption Capabilities:* Strong decryption for products like Dell Data Protection, Symantec, McAfee.[30]
- *Automated Processes:* Industry-leading processing capabilities for evidence preparation automation.
- *Investigative Abilities:* Built for investigators to compile and analyze evidence efficiently.
- *Reporting:* Provides flexible reporting framework for tailored case reports.

Figure 4.11 Encase Logo.

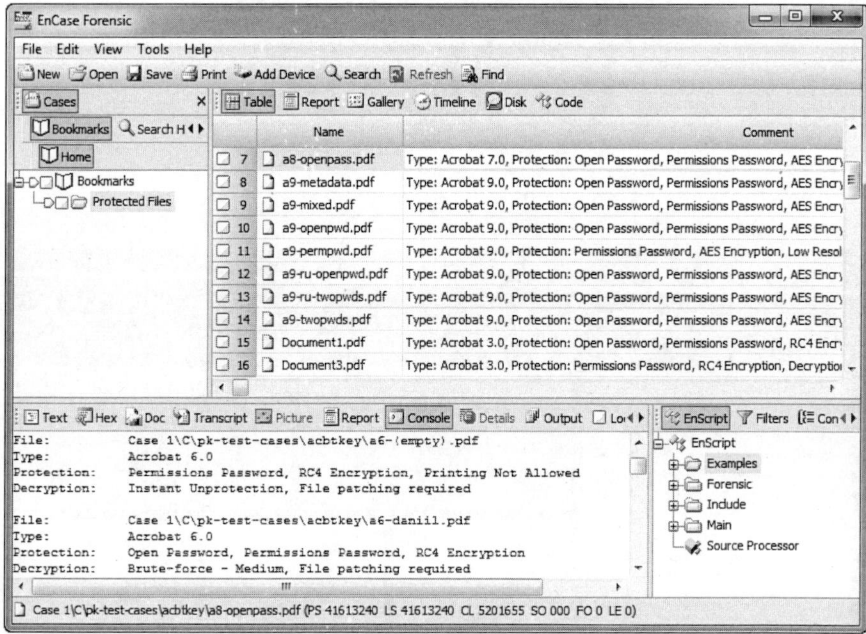

Figure 4.12 Encase Tool Interface.[31]

30 Image received from https://www.google.com/search?q=encase&newwindow=1&sca_esv=598080030&rlz=1C1GCEU_enIN1049IN1049&tbm=isch&sxsrf=ACQVn0-RNWXtx4I73n1OiMABnvogV2MmWQ:1705133036552&source=lnms&sa=X&ved=2ahUKEwiPxsDH89mDAxVOTWwGHbSiCooQ_AUoAXoECAQQAw&biw=1366&bih=599&dpr=1#imgrc=b0nBewRde53Q8M, visited on 13 January, 2024.

31 Image Retrieved from: https://support.passware.com/hc/en-us/articles/221742468-How-to-use-Passware-Kit-Forensic-with-Guidance-Software-EnCase.

(e) **Wireshark**

- *Functionality:* Open-source software for capturing, filtering, and visualizing data packets.
- *Key Functions:* Capture, filter, and visualize data packets for monitoring network communication.
- *Use Cases:* Government agencies, educational institutions, and corporates for network analysis.[32]

Figure 4.13 Wireshark Logo.

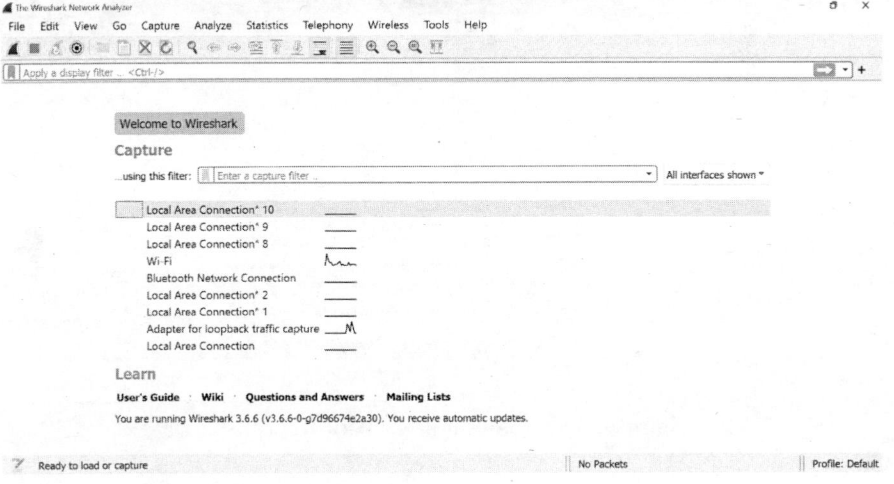

Figure 4.14 Wireshark Tool Interface.

(f) **MailXaminer**

- *Focus:* Email forensics tool supporting over 80 web and application-based email clients.

---

32 Image Received from: https://www.google.com/search?q=wireshark&tbm=isch&ved=2ahUKEwjXoezI89mDAxVIq2MGHejiD8gQ2-cCegQIABAA&oq=wireshark&gs_lcp=CgNpbWcQAzIFCAAQgAQyBQgAEIAEMgUIABCABDIFCAAQgAQyBQgAEIAEMgUIABCABDIFCAAQgAQyBQgAEIAEMgUIABCABDIFCAAQgAQ6BAgjECc6CggAEIAEEIoFEEM6BwgjEOoCECc6CAgAEIAEELEDOgQIABADUMEDWJovYNozaAFwAHgEgAGDAogBxRSSAQYwLjEzLjKYAQCgAQGqAQtnd3Mtd2l6LWltZ7ABCsABAQ&sclient=img&ei=70OiZZfuFcJWjuMP6MW_wAw&bih=599& biw=1366&rlz=1C1GCEU_enIN1049IN1049#imgrc=613tIhbCevprRM, visited on 13 January 2024.

- *Benefits:* Supports extensive email formats, offers flexible export options, and facilitates team collaboration.
- *Keyword Search:* Agile keyword search tool for instant evidence retrieval.
- *Usability:* User-friendly, enabling non-technical users to perform advanced email investigations.

Figure 4.15 MailXaminer Logo.

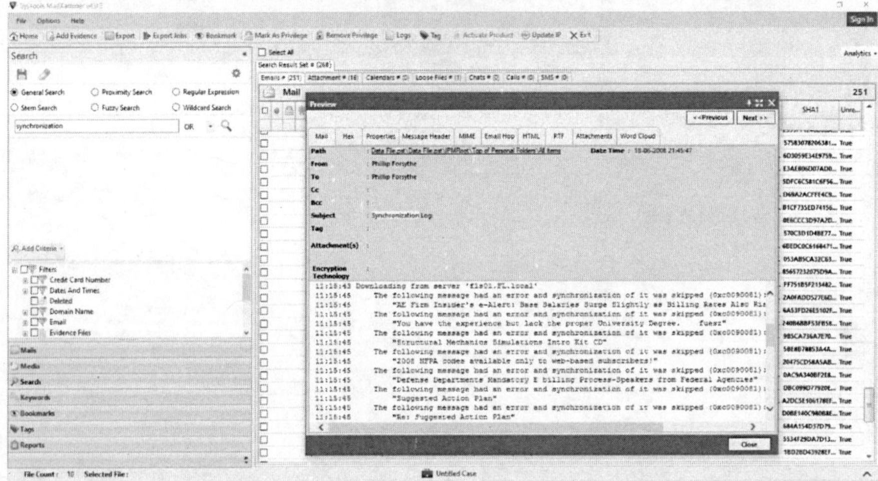

Figure 4.16 MailXaminer Tool Interface.[33]

(g) **FTK Imager**

- *Associated with FTK:* Part of the Forensic Toolkit (FTK) for computer forensics.
- *Disk Imaging:* Creates disk images in various formats, calculates hash values for integrity verification.
- *Wizard-Driven Approach:* Simplifies the detection of cybercrime.
- *Data Visualization:* Offers better visualization of data using charts.

33 Image Retrieved from: https://www.softwareadvice.com.sg/software/239762/mailxaminer.

- *Password Recovery:* Recovers passwords from over 100 applications.
- *Automated Analysis:* Provides advanced and automated data analysis features.

Figure 4.17 FTK Imager Logo.

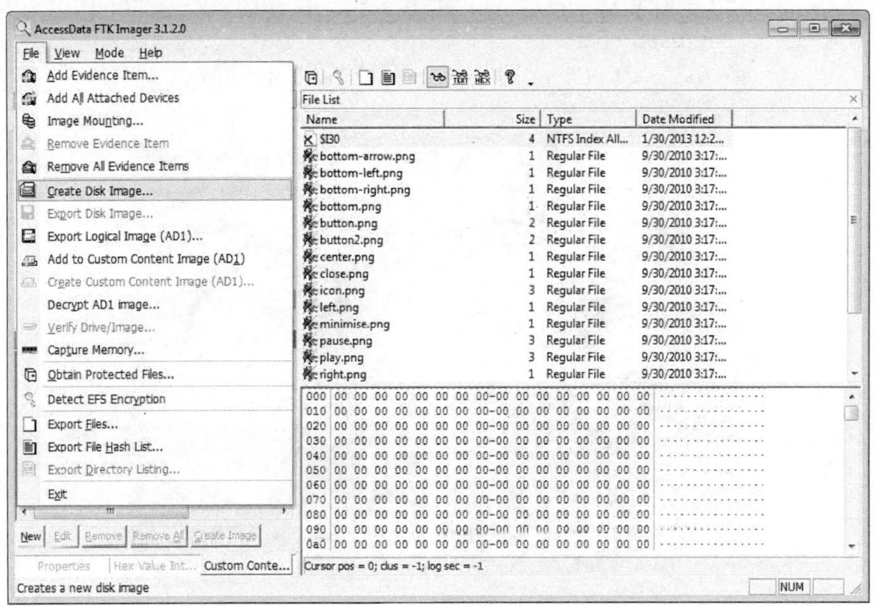

Figure 4.18 FTK Imager Tool Interface.

These tools collectively contribute to the forensic landscape, offering specialized capabilities for different aspects of digital investigations.

4.6 TYPES AND CATEGORIZATION OF DIGITAL FORENSICS

Digital forensics involves the collection, analysis, and preservation of digital evidence to investigate and prevent cybercrimes. It encompasses various subfields, each focusing on specific aspects of digital investigations, such as disk forensics, network forensics, web browser forensics, malware forensics, database forensics, mobile forensics and email forensics, etc.[34]

[A] Disk Forensics

- Disk forensics is akin to a digital archaeologist meticulously excavating information from storage media. Imagine a hard drive as a vast archaeological site—active files are the artifacts currently in use, deleted files are the buried relics, and unused regions are the unexplored territories. This discipline involves carefully uncovering these data layers to reconstruct the digital history.[35]
- *Example:* Analysing a hard drive might involve recovering deleted files, such as an important document thought to be lost, or identifying remnants of software installations in unused sectors.
- *Tools:* EnCase, FTK (Forensic Toolkit), Autopsy

Figure 4.19 Extracting Data from a Disk using a Tool.[36]

34 Simplified Guide to Digital Forensics available at: https://www.forensicsciencesimplified.org/digital/DigitalEvidence.pdf, visited on 12 January 2024.

35 Ambhore, P., Wankhade, A., & Meshram, B. B. (2018). Disk based Forensics Analysis.

36 Image received from: https://www.google.com/search?q=+Extracting+Data+from+a+Disk+using+a+tool&tbm=isch&ved=2ahUKEwjLzuOnn9eDAxU1TGwGHaY6BuMQ2-cCegQIABAA&oq=+Extracting+Data+from+a+Disk+using+a+tool&gs_lcp=CgNpbWcQAzoECCMQJ1CUA1j9B2CEDWgAcAB4AIAB4gGIAdcJkgEFMC4yLjSYAQCgAQGqAQtnd3Mtd2l6LWltZ8ABAQ&sclient=img&ei=Jd-gZYuoOrWYseMPpvWYmA4&bih=599&biw=1366&rlz=1C1GCEU_enIN1049IN1049#imgrc=Duj0wYzr3KmraM visited on 12 January 2024.

[B] Network Forensics[37]

- Network forensics is like being a digital detective overseeing a bustling city's traffic. The methods, "Catch it as you can" and "Stop, look, and listen," are analogous to traffic surveillance cameras capturing every movement. Tools like Wireshark act as detective lenses, allowing professionals to analyze the flow of data and discern patterns or anomalies.[38]
- *Example:* Investigating a cybersecurity breach might involve tracing the network traffic to identify the entry point and the data accessed by the intruder.
- *Tools:* Wireshark, Snort, Tcpdump.

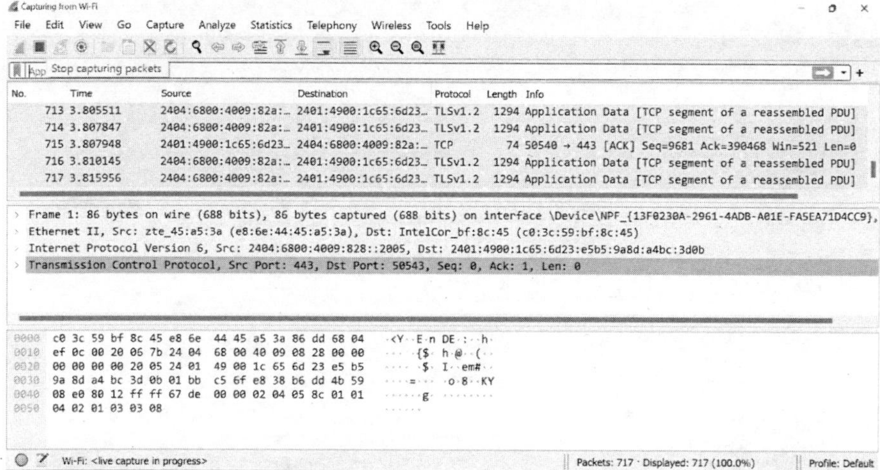

Figure 4.20 Investigating a Network Traffic using Wireshark Tool.[39]

[C] Web Browser Forensics

- Web browser forensics resembles investigating a personal diary, where login credentials, downloaded data, and

37 Simplified Guide to Digital Forensics available at: https://www.forensicsciencesimplified.org/digital/DigitalEvidence.pdf, visited on 12 January 2024.

38 Khan, S., Gani, A., Wahab, A. W. A., Shiraz, M., & Ahmad, I. (2016). Network forensics: Review, taxonomy, and open challenges. *Journal of Network and Computer Applications*, 66, 214–235.

39 Image received from: https://www.google.com/search?q=Investigating +a+network +traffic+using+Wireshark+Tool&newwindow=1&sca_esv=597734025&rlz=1C1GCEU_enIN1049IN1049&tbm=isch&sxsrf=ACQVn09MLkYkTccEt8j950kyd8zxIuJp8w:1705041815460&source=lnms&sa=X&ved=2ahUKEwiBjfPdn9eDΛxXnZWwGHY0DDfEQ_AUoAnoECAQQBA&biw=1366&bih=599&dpr=1#imgrc=aFj8gtCtUMoVvM, visited on 12 January 2024.

navigation history are entries reflecting user activities. Think of browser artifacts as the footprints left in the diary, providing insights into digital journeys.

- *Example:* Discovering login credentials and website visits can be crucial in understanding an individual's online behavior during an investigation.[40]
- *Tools:* Browser history tools, SQLite DB Browser, Forensic Browser.

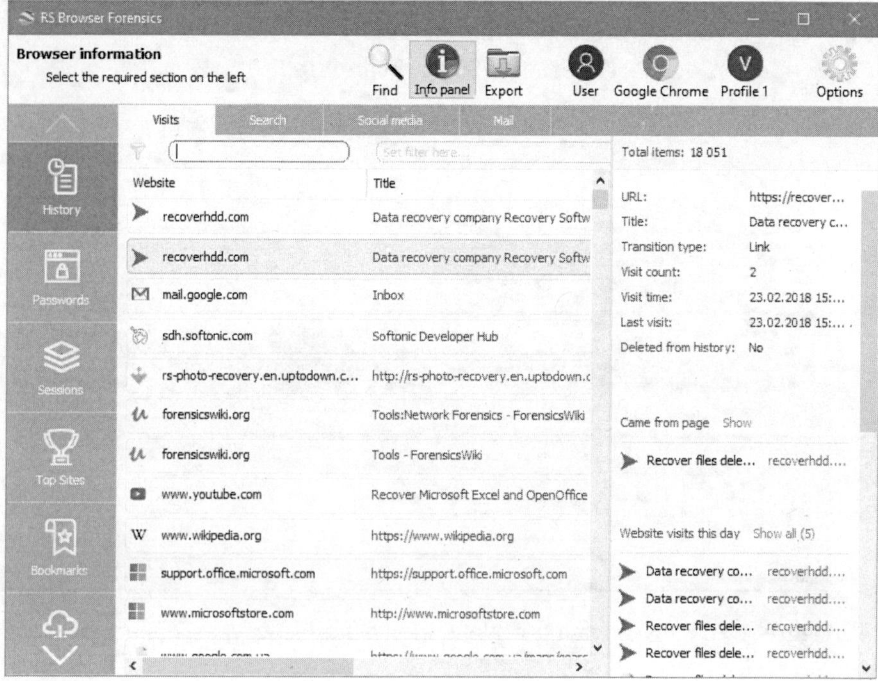

Figure 4.21 Investigating a Web Browser using a Browser Forensic Tool.[41]

40 Nelson, R., Shukla, A., & Smith, C. (2020). Web Browser Forensics in Google Chrome, Mozilla Firefox, and the Tor Browser Bundle. *Digital Forensic Education: An Experiential Learning Approach*, 219–241.

41 Image received from: https://www.google.com/search?q=Investigating+a+web+browser+using+a+Browser+Forensic+tool&newwindow=1&sca_esv=597734025&rlz=1C1GCEU_enIN1049IN1049&tbm=isch&sxsrf=ACQVn09tsvYiUQHBIW0AuD8gN6ONhn-OXw:1705041871058&source=lnms&sa=X&ved=2ahUKEwiirrT4n9eDAxXnR2wGHe27BlwQ_AUoA3oECAEQBQ&biw=1366&bih=599&dpr=1#imgrc=lVzd64HWAlDKMM, visited on 12 January 2024.

[D] Malware Forensics[42]

- Malware forensics is the pursuit of understanding a digital virus's origin, methods, and impact—an investigation into the 'patient zero' of the digital world. Static analysis is like studying the virus's blueprint, dynamic analysis is observing its behaviour in a controlled environment, and hybrid analysis combines both approaches.
- *Example:* Analyzing a ransomware attack might involve static analysis to identify the encryption algorithm used and dynamic analysis to observe its impact on files.
- *Tools:* Browser history tools, SQLite DB Browser, Forensic Browser.[43]

Figure 4.22 Malware Forensics.[44]

42 Introduction to Digital Forensics, available at https://courseware.cutm.ac.in/wp-content/uploads/2020/06/DF-1.pdf, visited on 13 January, 2024.

43 Malware Forensics: Discovery of the Intent of Deception y of the Intent of Deception, *Journal of Digital Forensics, Security & Law*, available at: https://commons.erau.edu/cgi/viewcontent.cgi?article=1082&context=jdfsl, visited on 12 January 2024.

44 Image received from: https://www.google.com/search?q=malware+forensics&newwindow=1&sca_esv=597734025&rlz=1C1GCEU_enIN1049IN1049&tbm=isch&sxsrf=ACQVn09pT-ZIEF-b3wxDOrdA4BeKy_GdCg:1705041962216&source=lnms&sa=X&sqi=2&ved=2ahUKEwjjqfCjoNeDAxX0SmwGHSf0DyEQ_AUoAXoECAYQAw&biw=1366&bih=599&dpr=1#imgrc=mlREZpEOalgkZM, visited on 12 January 2024.

[E] Database Forensics[45]

- Database forensics is akin to scrutinizing a company's ledger. It involves examining timestamps on database entries, ensuring financial transactions (or data manipulations) are accounted for and validating the actions of 'database users'—the bookkeepers of the digital realm.
- *Example:* Investigating a data breach might involve checking the database timestamps to trace when and how unauthorized access occurred.
- *Tools:* DB Browser for SQLite, Oracle Forensic Toolkit, Magnet AXIOM.

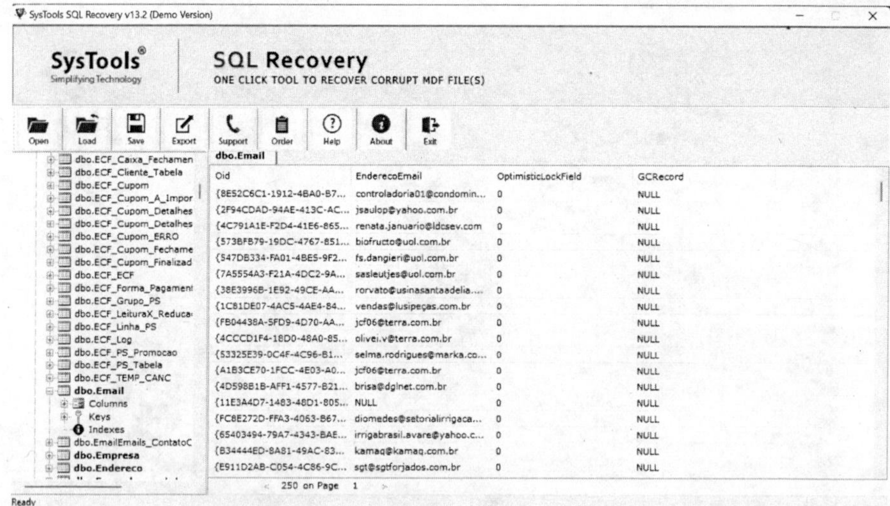

Figure 4.23 Investigating a Database using SysTools SQL Recovery Tool.[46]

[F] Mobile Forensics

- Mobile device forensics is like exploring a personal pocket universe—analyzing call logs, paired device history, and multimedia. The mobile device becomes a treasure trove of information, revealing a person's digital footprint.

45 Categories of Digital Forensics available at: https://ec.europa.eu/programmes/erasmus-plus/project-result-content/2a54509d-b6bb-43d8-8250-eae26782c392/FORC%20Book%201.pdf, visited on 13 January, 2024.

46 Image Received from: https://www.google.com/search?q=SysTools+SQL +Recovery +tool&newwindow=1&sca_esv=597734025&rlz=1C1GCEU_enIN1049IN1049 &tbm=isch&sxsrf=ACQVn09pol_AKAQUTA3Huoks21eeja 9fBw:1705042695148 &source=lnms&sa=X&ved=2ahUKEwj57a6Bo9eDAxXASGwGHZ7ED2IQ_AUoA3oECAUQBQ&biw=1366&bih=599&dpr=1#imgrc= BOMDgHzfoVzb4M, visited on 12 January, 2024.

- *Example:* Examining a suspect's smartphone might involve tracing their movements using location data or uncovering deleted messages relevant to an investigation.
- *Tools:* Cellebrite, Oxygen Forensic Detective, XRY.

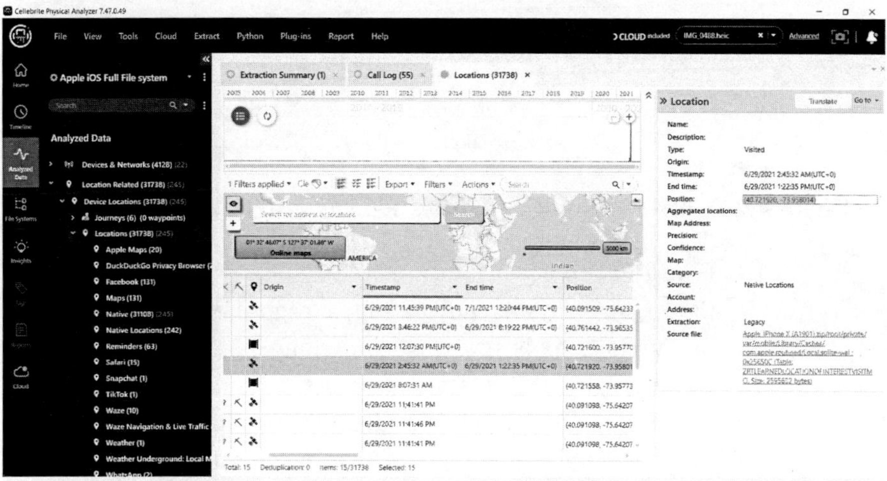

Figure 4.24 Investigating a Mobile Using Cellebrite Analyzer Tool.[47]

[G] GPS Forensics[48]

- GPS forensics is akin to studying a map with annotations. It involves examining GPS devices to retrieve track logs, points, and stored locations—a digital atlas of someone's journeys.
- *Example:* Investigating a missing person case might involve analyzing the GPS data from their device to reconstruct their movements.
- *Tools:* GPS Track Editor, Garmin BaseCamp, Forensic GPS Tools.

47 Image received from: https://www.google.com/search?q=Investigating+a+Mobile+using+Cellebrite+Analyzer+Tool&newwindow=1&sca_esv=597734025&rlz=1C1GCEU_enIN1049IN1049&tbm=isch&sxsrf=ACQVn08eLKQP4yRXpdrPNGEyb7MJlYCzJw:1705042635884&source=lnms&sa=X&ved=2ahUKEwju0Y3loteDAxXccGwGHaonAmgQ_AUoAnoECAQQBA&biw=1366&bih=599&dpr=1#imgrc=LaYAP8tBtCNT-M, visited on 12 January 2024.

48 Categories of Digital Forensics available at: https://ec.europa.eu/programmes/erasmus-plus/project-result-content/2a54509d-b6bb-43d8-8250-eae26782c392/FORC%20Book%201.pdf, visited on 13 January, 2024.

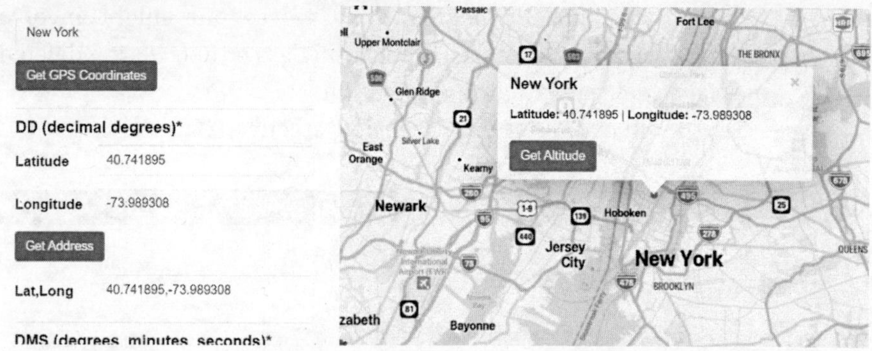

Figure 4.25 Investigating Coordinates Retrieved from a GPS Device.[49]

[H] Cloud Forensics

- Cloud forensics is navigating a digital sky. Professionals use various tools to investigate cloud platforms, ensuring a comprehensive understanding of activities in this virtual space.
- *Example:* Examining cloud storage might reveal shared files, timestamps, or user activities relevant to a corporate espionage investigation.
- *Tools:* Cloud Forensics Framework, Magnet AXIOM Cloud, Google Cloud Forensics.

Figure 4.26 Cloud Forensics.[50]

49 Image received from: https://www.google.com/search?q=Investigating+Coordinates+retrieved+from+a+GPS+device&newwindow=1&sca_esv=597734025&rlz=1C1GCEU_enIN1049IN1049&tbm=isch&sxsrf=ACQVn0_VG7DCzSKNnocoCve6E0tpAlNJ0Q:1705042569036&source=lnms&sa=X&ved=2ahUKEwj6153FoteDAxVbcWwGHZZsDdkQ_AUoAXoECAIQAw&biw=1366&bih=599&dpr=1#imgrc=DqJFq0ZxOZI16M, visited on 12 January, 2024.

50 Image received from: https://www.google.com/search?q=cloud+forensics&newwindow=1&sca_esv=597734025&rlz=1C1GCEU_enIN1049IN1049&tbm=isch&sxsrf=ACQVn0__4ORLLvdfqYNss0c93GqaUyjUhw:1705042518614&source=lnms&sa=X&sqi=2&ved=2ahUKEwjRlJitoteDAxWTdmwGHXj9A_8Q_AUoAXoECAQQAw&biw=1366&bih=599&dpr=1#imgrc= Krs1X2J4CdF1TM, visited on 12 January 2024.

[I] Email Forensics[51]

- Email forensics is like piecing together a digital conversation puzzle. Recovering deleted emails, calendar entries, and contacts paints a vivid picture of the parties involved and their interactions.
- *Example:* Uncovering a fraudulent activity might involve tracing communication patterns through email forensics.
- *Tools:* MailXaminer, Forensic Email Collector, Exif Pilot.

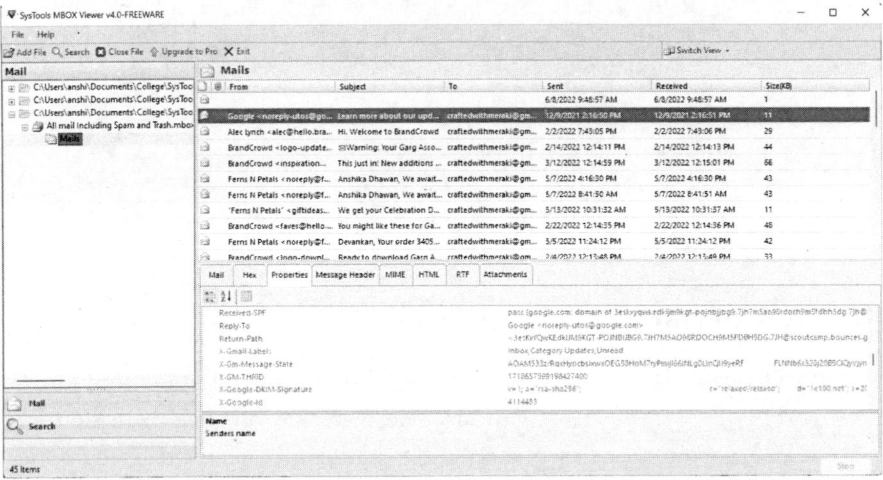

Figure 4.27 Investigating a Mailbox using SysTools MBOX Examiner Tool.[52]

[J] Memory Forensics

- Memory forensics is delving into the subconscious of a digital brain—extracting raw data from system memory. It's like exploring the mind of a computer to reveal its recent thoughts and activities.

51 Categories of Digital Forensics available at: https://ec.europa.eu/programmes/erasmus-plus/project-result-content/2a54509d-b6bb-43d8-8250-eae26782c392/FORC%20Book%201.pdf, visited on 13 January, 2024.

52 Image received from: https://www.google.com/search?q=Investigating+a+Mailbox+using+SysTools+MBOX+Examiner+Tool&newwindow=1&sca_esv=597734025&rlz=1C1GCEU_enIN1049IN1049&tbm=isch&sxsrf=ACQVn08GBJNVNWtkhEvL8vqT3Q1EGTc0zw:1705042453687&source=lnms&sa=X&ved=2ahUKEwi8pp2OoteDAxVBaGwGHdT5BgEQ_AUoAnoECAEQBA&biw=1366&bih=599&dpr=1#imgrc=_jYDyyatR9-djM, visited on 12 January 2024.

- *Example:* Investigating a cyberattack might involve extracting malware code from memory to understand its behavior and origin.
- *Tools:* Volatility, Rekall, WinDbg.[53]

Figure 4.28 Memory Forensics.[54]

4.7 ANTI-FORENSICS TOOLS

Anti-forensic tools are software or techniques designed to impede or thwart digital forensic investigations. The primary purpose of these tools is to erase or alter traces of activities that could be used as evidence in legal or investigative processes. While some anti-forensic tools are developed for legitimate purposes, such as protecting user privacy, they can also be exploited for illegal or malicious activities. These tools cover a broad spectrum of functionalities, aiming to counteract the techniques and tools used by digital forensic experts to recover and analyze electronic evidence. Anti-forensic methods are intended to shield cyber attackers from being identified. The complexity for digital forensic investigators has increased as a result of developments in anti-forensic techniques. There are a number of anti-forensic techniques, some of which are listed below:[55]

---

53 Categories of Digital Forensics available at: https://ec.europa.eu/programmes/erasmus-plus/project-result-content/2a54509d-b6bb-43d8-8250-eae26782c392/FORC%20Book%201.pdf, visited on 13 January, 2024.

54 Image received from https://www.google.com/search?q=memory+forensics&newwindow=1&sca_esv=597734025&rlz=1C1GCEU_enIN1049IN1049&tbm=isch&sxsrf=ACQVn09jqVvjnTCFD9ZKT29tnYaqMd1l6w:1705042404846&source=lnms&sa=X&sqi=2&ved=2ahUKEwjnlvj2odeDAxUGTmwGHdqlAAoQ_AUoAXo ECAYQAw&biw=1366&bih=599&dpr=1#imgrc= mk-kaeXjxM0eCM, visited on 12 January 2024.

55 Conlan, K., Baggili, I., & Breitinger, F. (2016). Anti-forensics: Furthering digital forensic science through a new extended, granular taxonomy. *Digital investigation*, 18, S66–S75.

(a) *Overwriting/Wiping:* It can also be referred to as secure erase. It is a procedure to overwrite data on a disk and make it unreadable. This way, the data is deleted permanently and cannot be recovered using any software or tool. There are numerous programs that can be used to overwrite important data on a storage device, making it difficult or impossible to recover. These applications have the ability to overwrite data and metadata.[56] Data Wiping Tools securely delete or overwrite data on storage media to make it unrecoverable. This can include file shredders or disk-wiping utilities.

(b) *File Hiding and Obfuscation Tools:* Tools that hide files, directories, or system artifacts from standard forensic analysis tools. This can include hiding files within alternate data streams or using techniques to manipulate file metadata.

(c) *Network Anti-Forensic Tools:* Techniques or tools that manipulate network traffic to hinder the detection of malicious activities. This may involve obfuscating communication patterns or using anonymity networks like Tor.

(d) *Rootkits and Malware:* Malicious software that is specifically designed to evade detection and removal by traditional antivirus and anti-malware tools. Rootkits, in particular, can manipulate system functions to conceal their presence.

(e) *Steganography:* It is a method to hide information within other files like images, video files, etc. in such a way that when you open the said file, it appears to be the same as it was before transformation. Various techniques may be implemented to achieve the same:

    (i) Hiding Multiple Files in an Image File
    (ii) Hiding Text Message in an Image File
    (iii) Hiding Image File in another Image File
    (iv) Hiding a Code behind an Image File

    and many more.

56 Wani, M. A., AlZahrani, A., & Bhat, W. A. (2020). File system anti-forensics–types, techniques and tools. *Computer Fraud & Security*, 2020(3), 14–19.

(f) *Encryption and Cryptography:* Encryption Tools are programs that encrypt data to make it inaccessible without the proper decryption key. While encryption itself is a legitimate method for protecting sensitive information, it can be employed as an anti-forensic measure when used to conceal criminal activities.[57] In order to make the data opaque to any attacker (or CFT) without the key, cryptographic file systems transparently encrypt data when it is written to the disk and decrypt data when it is read back. These file systems are currently easily accessible for Linux, Mac OS, and Windows. Although encryption has a very high entropy and that many products incorporate specific flags, headers, or other signatures into their encrypted data, encryption is very effective at concealing information. If the key is obtained, some forensic tools can decrypt encrypted data.[58]

(g) *General Data Obfuscation:* Data can also be concealed in unallocated or otherwise inaccessible areas that are overlooked by the current generation of forensic tools. Tail Obfuscation is one of its type. To deceive the investigator in this case, the attacker provided some false information (e.g. false email headers, changing file extensions). In light of this, the investigator may overlook information with forensic value. Another is the Covert Channel. A communication protocol known as a covert channel enables an attacker to get around intrusion detection systems and conceal data on a network. The attacker uses it to conceal his connection to the compromised system.

It's important to note that the use of anti-forensic tools for illicit purposes is generally considered illegal and unethical. In many jurisdictions, attempting to obstruct a legitimate digital forensic investigation can lead to legal consequences.[59] As digital forensic techniques and tools continue to advance, there is an ongoing cat-and-mouse game between forensic experts and those seeking to evade detection using anti-forensic measures.

---

57 De Beer, R., Stander, A., & Van Belle, J. P. (2015). Anti-forensics: A practitioner perspective. *Int. J.-Cyber-Secur. Digit. Forensics*, 4, 390–403.

58 https://www.researchgate.net/publication/228339244_Antiforensics_Techniques_detection_and_countermeasures.

59 Doe, J. (2021). Evading Detection: A Comprehensive Analysis of Anti-Forensic Tools. *Journal of Cybersecurity Research*, 7(2), 123–145.

A person versed with basic computer knowledge, or understands simple programming etc., may be able to do work easily. Other than this, there are a number of open-source free tools available that make the job easier and quicker.

LET'S RECALL

- Digital forensics, a subset of forensic science, focuses on recovering and investigating information from digital devices. It involves applying computer science and investigative procedures to examine digital evidence, ensuring adherence to legal protocols, validation with mathematics, and the use of validated tools.
- While often used interchangeably, these terms have distinct meanings. Digital forensics encompasses the recovery, preservation, analysis, and presentation of digital evidence. Computer forensics is a subset, focusing on computer systems. Cyber forensics deals specifically with cybercrimes, utilizing digital and internet technology.
- Digital forensics faces technical challenges like encryption and steganography, legal challenges including jurisdictional issues, and resource challenges such as the scarcity of experts. These challenges impact the collection and analysis of forensic media.
- Various tools support digital forensic investigations, categorized into data preservation, data recovery, data analysis, data reporting, and network utilities. Examples include FTK Imager, Recuva, Wireshark, and EnCase.
- The categorization of digital forensics into disk forensics, network forensics, web browser forensics, malware forensics, and others provides a structured understanding of its diverse applications. Each category, from examining storage media to analyzing network traffic and mobile devices, serves a unique investigative purpose.
- Anti-Forensic Techniques aim to shield cyberattackers. Techniques include overwriting/wiping, steganography, encryption, general data obfuscation, tail obfuscation, and covert channels. Each technique adds complexity to digital forensic investigations.

A. CHOOSE THE CORRECT OPTION

1. What does GPS Forensics primarily involve?
   (a) Analyzing browser history
   (b) Studying database entries
   (c) Retrieving track logs and stored locations
   (d) Investigating mobile phones
2. What category does Disk Forensics fall under?
   (a) Hardware Forensics
   (b) Network Forensics
   (c) Memory Forensics
   (d) Data Recovery
3. Which technique involves hiding information within other files like images or videos?
   (a) Data Recovery
   (b) Steganography
   (c) Encryption
   (d) General Data Obfuscation
4. What tool is commonly used for capturing and analyzing network packets in Network Forensics?
   (a) Autopsy
   (b) Wireshark
   (c) FTK Imager
   (d) EnCase
5. Which anti-forensic technique involves overwriting data on a disk to make it permanently unreadable?
   (a) Steganography
   (b) Encryption
   (c) Overwriting/Wiping
   (d) General Data Obfuscation

Answer Key

1. (c) Retrieving track logs and stored locations
2. (a) Hardware Forensics
3. (b) Steganography
4. (b) Wireshark
5. (c) Overwriting/Wiping

B. ANSWER THE FOLLOWING

1. Illustrate the working and key features of Autopsy, X-Ways Forensics, Cellebrite UFED, EnCase, Wireshark, and MailXaminer. How do these tools cater to different aspects of digital investigations?
2. Distinguish between Digital Forensics, Computer Forensics, and Cyber Forensics. How do they complement each other in investigative processes?
3. Explore the categorization of Digital Forensics tools. How do data preservation, recovery, analysis, and reporting tools contribute to the investigative process?
4. Discuss the challenges faced in the field of Digital Forensics, categorizing them into technical, legal, and resource-related issues.
5. Define Digital Forensics and explain its significance in modern investigations.

C. THINK OUTSIDE THE BOX

1. Delve into the world of anti-forensic techniques. How do strategies like overwriting, steganography, and encryption pose challenges to digital forensic investigators?

CHAPTER 5

Investigation in Digital Devices Navigating Digital Clues in Digital Forensics Investigations

A Glimpse into the Chapter

- Investigation in Digital Forensics: An Introduction
- Stages of Digital Forensics Investigation
- Pre-Investigation Phase
- During the Investigation
- Post-Investigation Stage
- Cardinal Rules for Investigation of Digital Devices

Digital device investigation necessitates a comprehensive strategy that includes legal knowledge, ethical considerations, and technological know-how. To ensure responsible and legal practices, a strong framework for digital investigations must develop as society continues to navigate the digital landscape. The interaction of difficulties, approaches, and moral issues highlights how important and complex investigations are in the digital era. In this chapter a detailed investigation procedure involved in digital devices is explained with images including Pre-Investigation, During-investigation and Post-Investigation. The relevance of this chapter lies in the discussion on the Cardinal Rules for the investigation of digital devices also.

5.1 INVESTIGATION IN DIGITAL FORENSICS: AN INTRODUCTION

Investigation in digital devices refers to the systematic process of collecting, analyzing, and preserving electronic evidence to uncover facts related to cybercrimes, security incidents, or legal cases. It involves a series of steps to ensure that digital evidence is handled appropriately and can be presented effectively in legal proceedings.[1] Investigation in digital devices demands a holistic approach encompassing technological expertise, ethical considerations, and legal awareness. As society continues to navigate the digital landscape, a robust framework for digital investigations must evolve to ensure responsible and lawful practices. The interplay of challenges, methodologies, and ethical considerations underscores the complexity and significance of investigations in the digital age.[2] Investigating digital devices poses unique challenges, ranging from the volatile nature of electronic evidence to data encryption. Challenges such as data privacy concerns, evolving technologies, and the prevalence of anti-forensic tools demand a sophisticated approach to digital investigations.[3] Therefore, as a student and practitioner, it is essential to understand the investigation of digital devices in detail.

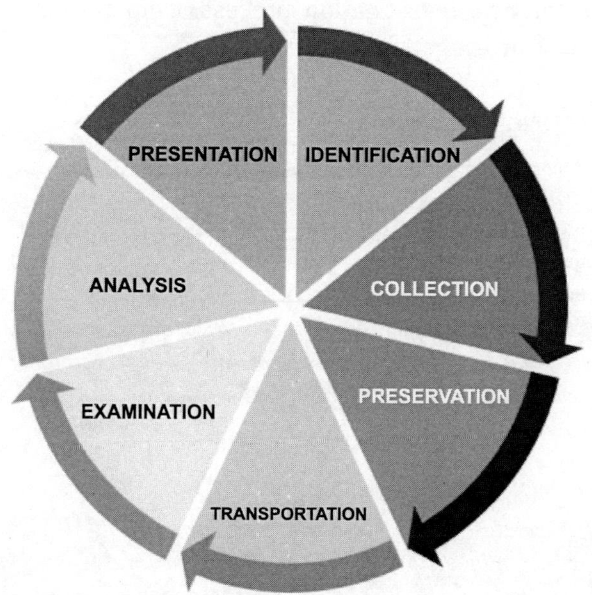

Figure 5.1 Stages of a Digital Forensic Investigation.

1 Casey, E. (2009). *Handbook of digital forensics and investigation*. Academic Press.

2 Johnson, C. D., & Smith, R. W. (2020). Ethics in Digital Investigations: Balancing Technological Advancements and Individual Rights. *Journal of Digital Ethics*, 8(1), 45–62.

3 Smith, J.A., & Brown, M. R. (2019). Challenges in Digital Investigations: A Comprehensive Analysis. *Journal of Cybersecurity Research*, 15(2), 87–104.

In digital forensics investigations, there are three primary stages[4]: [1] Pre-investigation; [2] During-investigation; and [3] Post-investigation. These stages are crucial for ensuring the success and integrity of the investigation process.

Connecting the various stages of digital forensic investigations, including the identification, collection, preservation, transportation, analysis, and presentation of digital evidence, to the pre-investigation, during investigation, and post-investigation stages can provide a comprehensive view of how these activities flow together.

Each stage plays a unique role in the overall investigative process, with the pre-investigation stage setting the groundwork, during-investigation stage being the active data collection and analysis phase, and the post-investigation stage involving reporting and legal proceedings. This comprehensive approach ensures that digital evidence is properly identified, collected, preserved, analyzed, and presented throughout the entire digital forensic investigation process, from the pre-investigation stage to the post-investigation stage.[5]

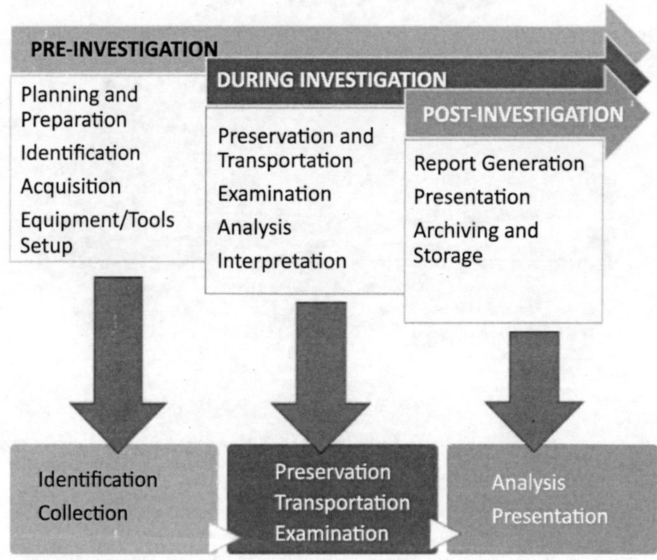

Figure 5.2 Overview of Pre, During and Post-Investigation Stages.

4 Pande, J., & Prasad, A. (2016). Digital forensics. Uttarakhand Open University. Also see: Jeetendra Pande and Ajay Prasad, Digital Forensics, Post-Graduate Diploma in Cyber Security Digital Forensics (PGDCS-06) retrieved from: https://uou.ac.in/sites/default/files/slm/MIT(CS)-202.pdf, visited on 13 January 2024.

5 Alharbi, S., Weber-Jahnke, J., & Traore, I. (2011). The proactive and reactive digital forensics investigation process: A systematic literature review. In *Information Security and Assurance: International Conference,* ISA 2011, Brno, Czech Republic, August 15–17, 2011. *Proceedings* (pp. 87–100). Springer Berlin Heidelberg.

5.2 SETTING THE STAGE: THE PRE-INVESTIGATION PHASE

The pre-investigation stage in digital forensics, often referred to as the "preliminary investigation" or "pre-forensic analysis," is a critical phase in the digital forensic process. It lays the foundation for a successful investigation and helps forensic experts determine the scope and objectives of the case. In this phase, investigators gather information, set objectives, allocate resources, assess risks, address legal considerations, and establish a solid foundation for the digital forensic investigation.[6] A well-executed pre-investigation significantly enhances the chances of a successful outcome.

5.2.1 Planning and Preparation

Preparing for the investigation is much more important than the primary investigation process. To understand this, let us assume that a student has to appear for an exam; now, before appearing for the same the student needs to be prepared to appear for it. The student must be well-read, carry required stationery, know the rules and requirements (like the requirement of documents: ID card), etc. Similarly, investigators must be prepared to investigate the crime scene with the required skills, tools, and equipment. Following is the checklist for the same:[7]

- Define the scope and objectives of the investigation. Understand the nature of the case and the specific questions that need to be answered.
- Assemble a team of digital forensics experts with the necessary skills and tools.
- Obtain the required legal permissions and warrants for evidence collection.
- Establish a clear chain of custody protocol to track evidence from the beginning.

Tools and Equipment

To gather electronic evidence, a few basic instruments and equipment are needed. However, technological advancements could necessitate

6 Ieong, R.S. (2006). FORZA–Digital forensics investigation framework that incorporate legal issues. *Digital investigation*, 3, 29–36.

7 Khan, A., Wiil, U.K., & Memon, N. (2010, May). Digital forensics and crime investigation: Legal issues in prosecution at national level. *In 2010 Fifth IEEE International Workshop on Systematic Approaches to Digital Forensic Engineering* (pp. 133–140).

modifications in the necessary tools and equipment.[8] As a result, preparations should be undertaken to acquire the tools needed to gather electronic evidence. Cameras, notepads, sketch pads, evidence forms, crime scene tape, and markers are standard materials used in crime scene processing.

Tools and equipment are required for each stage of the operation (Documentation, collecting, packing, and transportation).[9]

1. **Documentation Tools**
   (a) Cable tags
   (b) Indelible felt-tip markers
   (c) Stick-on labels

Cable Tags

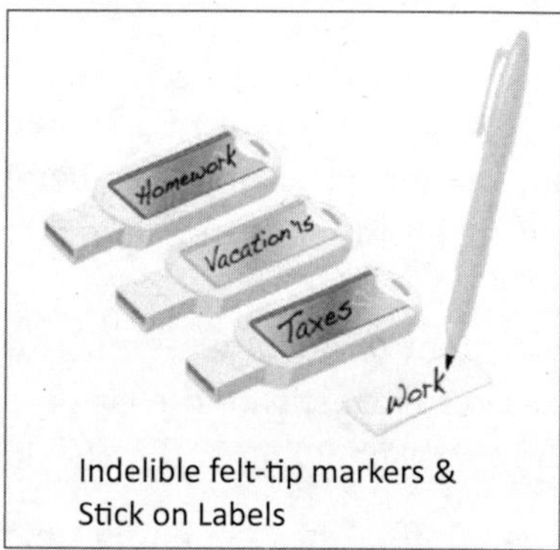

Indelible felt-tip markers & Stick on Labels

Figure 5.3 Pictorial Representation of Documentation Tools.

2. **Disassembly and Removal Tools**
   (a) Flat-blade and cross-tip screwdrivers
   (b) Hex-nut and secure-bit drivers
   (c) Star-type nut drivers
   (d) Needle-nose and standard
   (e) Small tweezers
   (f) Specialized screwdrivers (manufacturer specific)
   (g) Wire cutters

8 Chaubey, R. K. (2009). An Introduction to Cyber Crime and Cyber Law. Kamal Law House. Also see: Walden, I. (2007). *Computer crimes and digital investigations.* Oxford University Press.

9 Marshall, A. M. (2009). *Digital forensics: digital evidence in criminal investigations.* John Wiley & Sons.

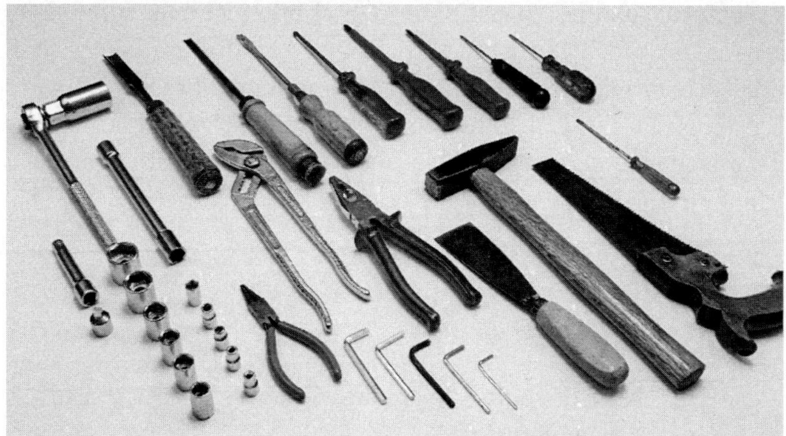

Figure 5.4 Disassembly and Removal Tools.

3. **Packaging and Transportation Tools**
   (a) Antistatic bags and bubble wrap.
   (b) Cable ties and evidence bags.
   (c) Evidence and packing tape.
   (d) Sturdy boxes of various sizes.
   (e) Faraday bags to pack mobile/wireless devices.

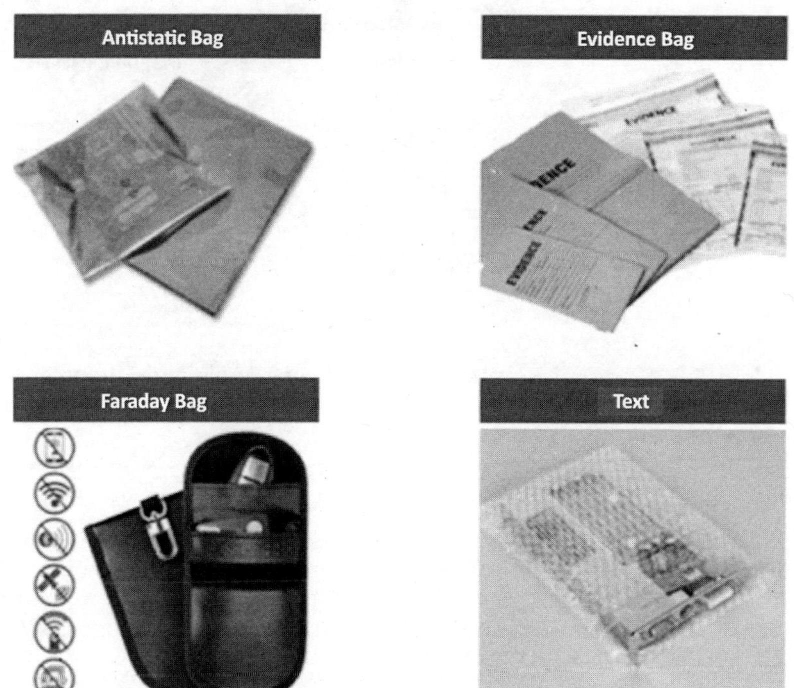

Figure 5.5 Pictorial Representation of Packaging and Transportation Tools.

When cell phones, smart phones, and other mobile communication devices are seized, investigators should have radio frequency shielding materials like faraday isolation bags or aluminium foil ready to wrap them in. A call, text message, or other communications signal that may tamper with the evidence cannot be received by the phones if they are covered with radio frequency shielding material.[10]

4. Other Tools and Equipment

(a) Evidence tags
(b) Evidence tape
(c) Gloves
(d) A list of contact telephone numbers for assistance
(e) A magnifying glass
(f) Printer paper
(g) A seizure disk
(h) A small flashlight

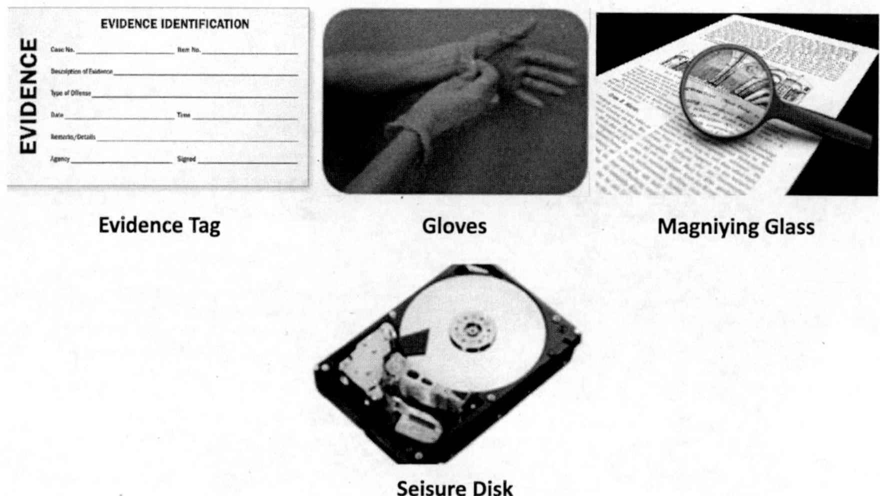

Figure 5.6 Pictorial Representation of Other Tools and Equipment.

10 Chaubey, R. K. (2009). An Introduction to Cyber Crime and Cyber Law. Kamal Law House. Also see: Irons, A., & Lallie, H. S. (2014). Digital forensics to intelligent forensics. *Future Internet,* 6(3), 584–596. Also see: https://www.justice.gov/sites/default/files/usao/legacy/2008/02/04/usab5601.pdf.

Figure 5.7 Categorized List of Tools and Equipment.

Apart from the content stated above, the investigation team must be equipped with a cyber forensic lab setup that includes all the necessary tools, computer hardware and software, etc.[11]

When embarking on an investigation, it is imperative to acknowledge that it involves more than just assembling the necessary tools and equipment. The process of planning and preparing extends beyond the mere collection of tools and equipment. It encompasses a comprehensive comprehension of specific rules, principles, and Standard Operating Procedures (SOPs) essential for successfully executing the investigative process. These rules and principles serve as the guiding framework for the investigation, ensuring that it is conducted systematically and lawfully.[12] They dictate the ethical and legal boundaries that investigators must adhere to, guaranteeing that the process is conducted with integrity and fairness. Moreover, the SOPs provide a structured set of instructions that outline the step-by-step procedures to be followed during the investigation. They help standardize the process, ensuring consistency and effectiveness in pursuing the investigative objectives.[13]

11 Cyber Crime Investigation Manual-Jharkhand Police. Retrieved from: https://jhpolice.gov.in/sites/default/files/documents-reports/jhpolice_cyber_crime_investigation_manual.pdf.

12 Common Phases of Computer Forensics Investigation Models retrieved from: https://www.researchgate.net/publication/228752802_Common_Phases_of_Computer_Forensics_Investigation_Models, visited on 13 January 2024.

13 Computer Forensics, retrieved from https://epgp.inflibnet.ac.in/epgpdata/uploads/epgp_content/law/07._information_and_communication_technology_/12._computer_forensics/et/5747_et_12_et.pdf, visited on 13 January, 2024.

Standard Operation Procedure[14]

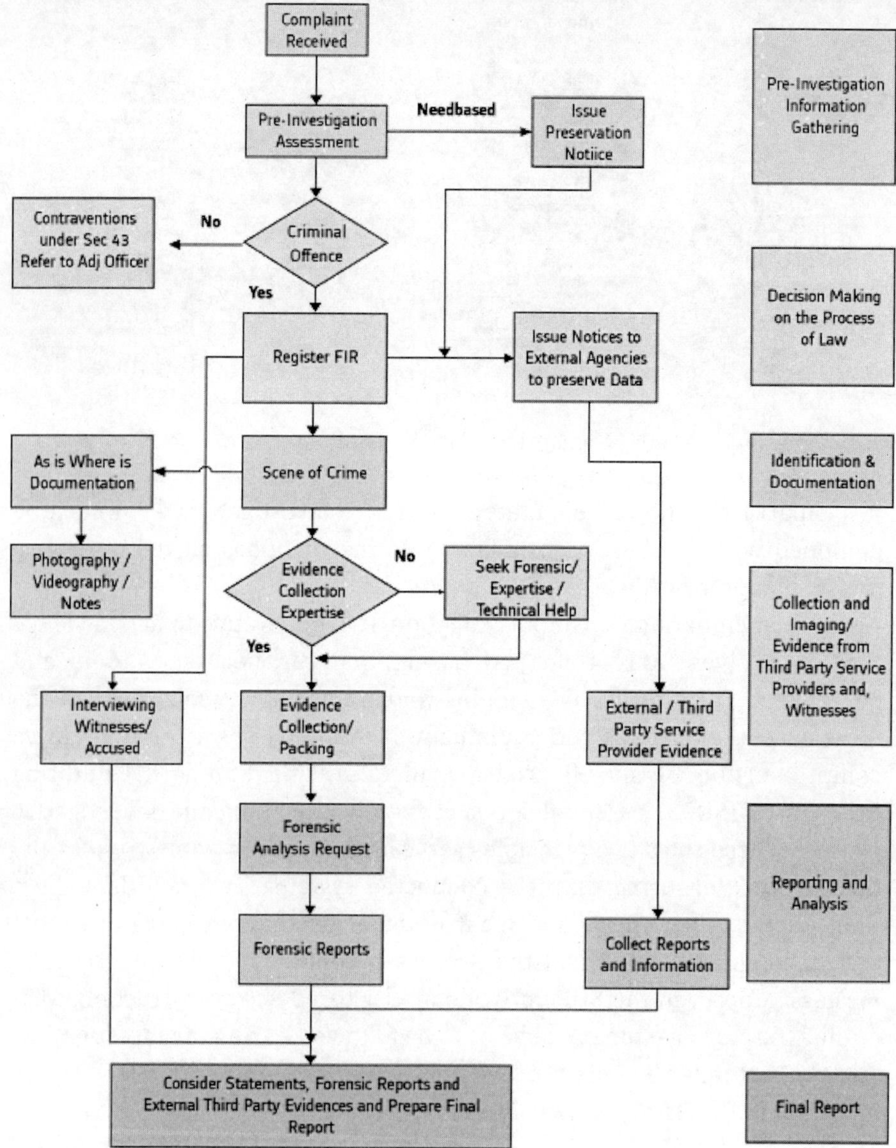

Figure 5.8 Flow Chart for Digital Crime Investigation under ITAA 2008[15].

---

14 Overview of Digital Forensics, retrieved from: https://informationsecurity.report/Resources/Whitepapers/9c6a1545-f876-48b6-9ca9-85f71e4d06cf_Overview-of-Digital-Forensics_whp_Eng_0315.pdf, visited on 14 January, 2024.

15 Retrieved from: Cyber Crime Investigation Manual–Jharkhand Police. https://jhpolice.gov.in/sites/default/files/documents-reports/jhpolice_cyber_crime_investigation_manual.pdf.

5.2.2 Identifying Potential Sources of Evidence

The foundation of a digital investigation is the identification phase. When an incident or a crime is identified, a hypothesis about what happened is formed. In terms of law, the identification phase entails examining the information infrastructure of the adversary and advising the lead attorney as to the location(s) of any potentially relevant evidence as well as the most effective method(s) of gathering it.[16]

Digital evidence can be found in a wide range of electronic devices and systems. The first step in evidence acquisition is to identify potential sources of digital evidence. This includes, but is not limited to, computers, mobile devices, servers, network logs, email servers, IoT devices, cameras, social media, external storage, GPS data, software logs, messaging apps, financial records, medical devices, educational systems, and automated sensors. These sources can yield important information for various types of investigations.

5.2.3 Acquisition

In general, there are two categories of data acquisition: static acquisitions and live acquisitions. Both methods and their data integrity requirements are similar. Data acquisition in digital forensics encompasses all the procedures involved in gathering digital evidence, including cloning and copying evidence from any electronic source. In general, there are two categories of data acquisition: static acquisitions and live acquisitions. Both methods and their data integrity requirements are similar.[17]

Live Data Acquisition

Live acquisition involves the collection of data from a running computer or device. This method is suitable for acquiring volatile data, information stored in the device's RAM (Random Access Memory). Volatile data is temporary and is lost when the device is powered off or rebooted.[18] Live

16 Wiles, J., & Reyes, A. (2007). The best damn cybercrime and Digital Forensics Book period. Syngress.

17 P. Stephenson, (2003). "A Comprehensive Approach to Digital Incident Investigation.", *Information Security Technical Report,* Vol. 8, Issue 2, pp 42–52. Also see: S. Ciardhuain, (2004) "An Extended Model of Cybercrime Investigation", *International Journal of Digital Evidence,* Vol. 3, No. 1, pp. 1–22.

18 F. C. Freiling & B. Schwittay, (2007). "Common Process Model for Incident and Computer Forensics", in *Proceedings of Conference on IT Incident Management and IT Forensics,* Stuttgard, Germany, pp. 19–40. Also see: S. Ciardhuain, (2004) "An Extended Model of Cybercrime Investigation", *International Journal of Digital Evidence,* Vol. 3, No. 1, pp. 1–22.

acquisition is employed when investigators need to capture real-time data, such as active processes, open files, system settings, and network connections. It is beneficial in cases where it's essential to capture information that might disappear when the device is shut down.

Static Data Acquisition

Static acquisition, however, focuses on non-volatile data stored on storage media like hard drives, solid-state drives, memory cards, or other persistent storage devices. This method is typically less intrusive and involves creating forensic images of the storage media. Static acquisition is ideal for capturing a complete and unaltered data snapshot on a storage device. It is used when investigators need to analyze files, directories, and data structures. Investigators can preserve the original evidence for examination without altering it by creating forensic images.[19]

Data Acquisition Methods

Digital forensics relies heavily on effective data acquisition to extract, preserve, and analyze electronic evidence. This note explores various data acquisition methods employed in digital forensics, highlighting the importance of choosing the appropriate method based on the nature of the investigation.[20] Broadly, Data Acquisition may be done in the following ways:[21]

1. **Disk-to-image File:** Disk-to-image method is considered the most common and flexible method for investigations. By using it, investigators can create several copies of a suspect's media which are constructed by using bit-for-bit replications mechanism. Moreover, they can also use other forensic tools, such as SMART, ProDiscover, X-Ways Forensics, FTK, ILook, and Autopsy, to read the most common types of disk-to-image files they created. These software tools treat the disk-to-image file as though it is the original disk.

---

19 Data Acquition Methods retrieved from: https://www.oreilly.com/library/view/practical-mobile-forensics/9781788839198/04ef437d-1dc9-439e-a228-7b7ea9c736e9.xhtml#:~:text=The%20following%20points%20break%20down,mobile%20forensics%20tool%20leveling%20system, visited on 16 January, 2024.

20 Legal Foundation for Electronic Evidence (2021). Legal Implications of Digital Investigations. Retrieved from https://www.legalfoundation.org/digital-investigation-legal, visited on 16 January, 2024.

21 Yusoff, Y., Ismail, R., & Hassan, Z. (2011). Common phases of computer forensics investigation models. *International Journal of Computer Science & Information Technology*, 3(3), 17–31.

2. **Disk-to-disk Copy:** Sometimes, investigators face problems of making a disk-to-image file. One of the main challenges is hardware and/or software incompatibilities. This becomes more complicated when investigators face old disk drives. In this case, they might have to create a disk-to-disk copy instead of disk-to-image copy. Several tools are available—like EnCase and SafeBack-that can copy data exactly from an older disk to a newer one. These tools can adapt their functionality to match data on the original suspect drive. They do that by modifying disk hardware features such as cylinder, head, and track configuration.[22]
3. **A Logical Disk-to-disk:** In the case that the extracted data is stored in a large drive, the data capture process can take several hours. To overcome this issue, some vendors suggest the use of sparse data acquisition that gathers only specific types of files.[23] In this case, the overall performance will be improved especially when the process needs to examine only some parts of the suspect's disk drive.
4. **A Sparse Copy of a Folder or File:** Some compression methods can be used to reduce file size to fit with the disk drive storage space. Standard tools for archiving such as WinRAR, PKZip and WinZip use lossless compression to reduce file size without affecting the image quality. One way to test a file's data consistency is to carry out a hashing method such as MD5 or SHA-1. This should be done before and after applying the compression. In this case, if the compression process is done correctly, both copies should have the same hashing value, otherwise, the compressed file is corrupted. Another method is to use a backup using tapes, such as Super Digital Linear Tape (SDLT) or Digital Audio Tape/Digital Data Storage (DAT/DDS), especially when working with large drives.

To determine the best acquisition method that can be used for digital investigation purposes, examiners must consider the following key points:

- The size of the data volume.
- The method of retaining the data as evidence or referring it to its owner.

22 Digital Forensics, retrieved from: https://www.eccouncil.org/cybersecurity/what-is-digital-forensics/, visited on 16 January 2024.

23 Asaro, T., & Biggar, H. (2007). Data de-duplication and disk-to-disk backup systems. Enterprise Strategy Group.

- The time required to carry out the acquisition.
- The location of the digital evidence.

Creating Forensic Images

Creating forensic images is a key component of evidence acquisition. A forensic image is an exact, bit-for-bit copy of the storage media, such as hard drives, solid-state drives, or mobile device memory. The creation of forensic images is important for several reasons:

1. **Preserving Integrity:** Forensic images preserve the original state of the evidence, including deleted files and unallocated space. This ensures that the data remains unaltered and intact throughout the investigation.
2. **Preventing Tampering:** Forensic images are write-protected, meaning no data can be written to the original storage media during imaging. This prevents any accidental or intentional changes to the evidence.
3. **Analysis and Examination:** Forensic images are used for analysis, allowing investigators to search for, recover, and analyze data without affecting the original evidence.

5.2.4 Documentation and Tool Setup

Documentation and equipment/tool setup are foundational elements of the digital forensics process.

Documentation

Document all pre-investigation activities, including the date and time of evidence collection, individuals involved, and the specific location of the evidence. At the outset of a digital forensic investigation, it is essential to document all activities that occur before the actual data acquisition and analysis. This includes recording the date and time of evidence collection, the names of individuals involved in the investigation, and the specific location where the evidence was collected. This Documentation establishes a clear timeline of events and ensures that all individuals involved in the investigation can be identified. It is crucial for maintaining the chain of custody, which is vital for legal proceedings. A well-documented chain of custody provides a reliable history of evidence handling. Moreover, it includes maintaining records of legal permissions, such as search warrants or subpoenas.

Equipment and Tool Setup

Setting up the necessary equipment and tools is a crucial step that directly impacts the effectiveness and integrity of a digital forensic investigation. Before examining digital evidence, it is imperative to verify that all forensic tools and equipment are in proper working condition. This includes hardware write-blockers, forensic imaging software, analysis tools, and any specialized hardware required for evidence acquisition. Functional tools and equipment are essential for ensuring that data is acquired accurately and without contamination. Using reliable and well-maintained tools helps prevent unintentional alterations to the evidence.

The workspace where digital evidence examination occurs must be clean, secure, and controlled. It should be free from contaminants and secure from unauthorized access. This includes establishing protocols for access control and maintaining a documented record of who enters the workspace. A clean and secure workspace minimizes the risk of tampering with the evidence during examination. Controlling access ensures that only authorized individuals are present, maintaining the integrity of the process.

5.3 DURING THE INVESTIGATION: UNRAVELING THE DIGITAL FORENSICS PROCESS

Digital forensics has emerged as a critical component of modern investigations, encompassing various activities to collect, analyze, and preserve digital evidence. The "during the investigation" stage is at the heart of digital forensics, where forensic experts leverage specialized tools and techniques to uncover critical information, reconstruct events, and build a case. In this in-depth exploration, we will delve into the intricacies of the during-investigation stage, shedding light on the methodologies, challenges, and best practices that guide this crucial phase of the digital forensics process.[24]

5.3.1 Preservation of Digital Evidence

Protecting and preserving the integrity of electronic data gathered during a digital forensic investigation is known as "preservation of digital evidence.".

24 Khan, A., Wiil, U. K., & Memon, N. (2010, May). Digital forensics and crime investigation: Legal issues in prosecution at national level. In *2010 Fifth IEEE International Workshop on Systematic Approaches to Digital Forensic Engineering* (pp. 133–140). IEEE.

It entails taking precautions to guarantee that the evidence is authentic, unaltered, and unchanged throughout the investigation. Maintaining digital evidence's admissibility in court is essential and enhances the investigation's overall credibility.[25] Preservation is a critical aspect of the during-investigation stage. Digital evidence can be fragile and easily tampered with. Forensic experts must ensure that the data remains unchanged during the investigation process. This involves employing write-blocking mechanisms to prevent any writing to the original evidence, thus preserving its integrity. Secure storage and chain of custody documentation are crucial for maintaining the evidentiary value.[26]

Before being packaged and delivered, every digital evidence must be inventoried, labelled, photographed, videotaped, or otherwise captured and documented before being packaged and delivered. All connections and associated devices should be labelled to make eventual system reconfiguration simple. It is important to note that digital evidence may contain latent, trace, or biological evidence, thus, suitable procedures must be taken to protect it. Accordingly, digital evidence imaging should be performed before performing latent, trace, or biological evidence processing on the evidence.

The following must be ensured while preserving Digital Evidence:

- Antistatic packing must be used for any digital evidence.
- Digital evidence should only be packaged using paper bags, envelopes, cardboard cartons, and antistatic containers.
- Plastic materials shouldn't be used when gathering digital evidence since they can generate or transmit static electricity and allow condensation and humidity to form, which might harm or destroy the evidence.
- The evidence must be shielded from damage such as bending, scratching, and other alterations.
- All packaging and storage containers used for digital evidence should have legible labels on them.
- Keep any cell phones, mobile devices, or smart phones in the on or off position they were discovered in.
- To prevent data messages from being broadcast or received by mobile or smart phone(s), wrap them in signal-blocking materials

25 Chaubey, R. K. (2009). *An Introduction to Cyber Crime and Cyber Law*. Kamal Law House.

26 Granja, F. M., & Rafael, G. D. R. (2017). The preservation of digital evidence and its admissibility in the court. *International Journal of Electronic Security and Digital Forensics*, 9(1), 1–18.

like faraday isolation bags, radio frequency shielding material, or aluminium foil.
- Additionally, grab all electrical connectors and power sources for any electronic gadgets.

5.3.2 Transportation of Digital Evidence

When digital evidence must be moved from the collection location to a forensic laboratory or secure storage location, it's transported under strict protocols. Any mishandling or improper transportation can compromise the integrity of the evidence.[27]

A clear chain of custody is established, documenting who handled the evidence at each stage to maintain its credibility in court.[28]

The following must be ensured while transporting digital evidence:

- It must be protected from magnetic fields created by speaker magnets, radio transmitters, and magnetic mount emergency lights.
- Do not store digital evidence in a car for an extended amount of time. Digital evidence can be harmed or destroyed by heat, cold, and humidity.
- Ensure that computers and electronic devices are packaged and secured to avoid shock and vibration damage during transportation.
- Keep track of the chain of custody for all transported evidence and document the movement of the digital evidence.

5.3.3 Examination of Digital Evidence

The heart of the during-investigation stage lies in the analysis of digital evidence. This phase involves extracting relevant information from the collected data. Forensic experts employ a range of tools and techniques to sift through vast amounts of data. They focus on identifying key information that can aid the investigation, such as documents, emails, chat logs, images, and metadata. Advanced techniques uncover hidden or deleted data that may hold critical clues.

---

27 Akilal, A., & Kechadi, M. T. (2021). A Forensic-Ready Intelligent Transportation System. In *International Summit Smart City 360°* (pp. 617–630). Cham: Springer International Publishing.

28 Granja, F. M., & Rafael, G. D. R. (2017). The preservation of digital evidence and its admissibility in the court. *International Journal of Electronic Security and Digital Forensics*, 9(1), 1–18.

The analysis of digital evidence is a multifaceted process that involves several steps and methodologies. Digital forensic experts employ techniques and specialized software tools to examine digital artefacts. Some of the critical aspects of digital evidence analysis include:[29]

1. **Data Recovery and Reconstruction:** Data loss is a common issue that affects individuals and businesses alike. Digital forensic experts use data recovery tools and techniques to retrieve lost information when data has been intentionally or accidentally deleted. They also reconstruct data to establish a timeline of events, uncover file changes, and determine user actions.[30]
2. **Keyword Searching:** Keyword searching is a fundamental aspect of digital evidence analysis. Investigators use relevant keywords and search patterns to identify specific information within vast data sets. This approach is valuable in quickly pinpointing critical evidence, such as documents or communications related to a case.[31]
3. **Metadata Examination:** Metadata, which provides essential information about files and their origins, is often a goldmine of evidence. Analyzing metadata can reveal valuable insights, such as when a file was created, modified, or accessed.[32]
4. **Timeline Analysis:** Creating a timeline of digital events is crucial for understanding the sequence of actions or communications relevant to an investigation. Timeline analysis helps investigators piece together the chronology of events, aiding in the reconstruction of the case.
5. **Data Carving:** Data carving is the process of extracting and reassembling fragmented or unallocated data. This technique

---

29 Digital Evidence and Forensics, retrieved from: https://nij.ojp.gov/digital-evidence-and-forensics, visited on 16 January, 2024.

30 Retrieved from: https://www.esecforte.com/services/data-recovery-reconstruction/, visited on 16 January, 2024.

31 Handling Digital Devices, retrieved from: https://www.unodc.org/e4j/zh/cybercrime/module-6/key-issues/handling-of-digital-evidence.html#:~:text=Analysis%20and%20Reporting,the%20analysis%20(%20reporting%20phase), visited on 16 January, 2024.

32 Handling Digital Devices, retrieved from: https://www.unodc.org/e4j/zh/cybercrime/module-6/key-issues/handling-of-digital-evidence.html#:~:text=Analysis%20and%20Reporting,the%20analysis%20(%20reporting%20phase), visited on 16 January, 2024. Also see: Cyber Security Coalition, Cyber Security Incident Management Guide, 2015.

is advantageous when dealing with damaged storage media or when critical information has been intentionally concealed.[33]

5.3.4 Challenges and Considerations during the Investigation

While the during-investigation stage is a vital component of the digital forensics process, it comes with several challenges and considerations:[34]

1. **Data Volume:** The digital world generates an immense amount of data daily. Sorting through this volume of data can be overwhelming. Forensic experts need tools and strategies to efficiently identify and extract relevant information without being bogged down by irrelevant data.[35]
2. **Data Encryption:** As encryption becomes more prevalent, accessing encrypted data poses a significant challenge. Digital forensic experts must employ decryption techniques or work with cryptographic experts to gain access legally and ethically.
3. **Privacy Concerns:** Collecting and analyzing digital evidence must be conducted to respect privacy and legal rights. Adhering to applicable laws and regulations is paramount, as improper data handling can lead to legal challenges and evidence exclusion.
4. **Cybersecurity Risks:** Handling digital evidence requires high security to prevent data breaches or tampering. The forensic laboratory's security and evidence storage are of utmost importance.
5. **Data Fragmentation:** Data can be fragmented and scattered across different devices and storage locations in digital environments. Reconstructing a comprehensive picture of events from these fragments can be complex and time-consuming.
6. **Emerging Technologies:** The fast-paced evolution of digital technologies presents ongoing challenges for forensic experts.

33 Handling Digital Devices, retrieved from: https://www.unodc.org/e4j/zh/cybercrime/module-6/key-issues/handling-of-digital-evidence.html#:~:text=Analysis%20and%20Reporting,the%20analysis%20(%20reporting%20phase), visited on 16 January, 2024.

34 Mughal, A. A. (2019). A Comprehensive Study of Practical Techniques and Methodologies in Incident-based Approaches for Cyber Forensics. *Tensorgate Journal of Sustainable Technology and Infrastructure for Developing Countries,* 2(1), 1–18.

35 Smith, J. A., & Brown, M. R. 2019. Challenges in Digital Investigations: A Comprehensive Analysis. *Journal of Cybersecurity Research,* 15(2), 87–104.

They must continuously update their knowledge and tools to keep up with new devices, applications, and data storage methods.

5.4 CLOSING THE CASE: THE POST-INVESTIGATION STAGE

The post-investigation stage in digital forensics is the phase that follows the active analysis of digital evidence during the investigation. It encompasses the final steps, including documenting and reporting findings, expert testimony, and handling digital evidence after the case's conclusion. This stage is crucial for ensuring the investigation's results are presented accurately and comply with legal requirements. Here, the details of the post-investigation stage are given as under:[36]

5.4.1 Key Aspects of the Post-Investigation Stage

1. **Documentation and Reporting:** Proper Documentation of the investigation process and findings is essential. Digital forensic experts must create comprehensive reports that detail the methods used, the evidence collected, and the analysis performed. These reports serve as an official record of the investigation and are often submitted in court as exhibits. These reports' clarity and thoroughness are vital for effectively presenting the findings.[37]
2. **Expert Testimony:** In many legal cases, digital forensic experts must provide expert testimony in court. This involves explaining the methods used in the investigation, the significance of the evidence, and the conclusions drawn. Testifying experts must convey complex technical information in a way that judges and jurors can understand.
3. **Legal Compliance:** Ensuring that all actions taken during the investigation, including the handling of digital evidence, comply with legal standards is paramount. This is especially important for maintaining the admissibility of evidence and preventing challenges from opposing parties.

36 Jeff Darington, "The Phases of the Digital Forensics Investigation Process" retrieved from: https://graylog.org/post/the-phases-of-the-digital-forensics-investigation-process/, visited on 16 January, 2024.

37 Cyber Crime Investigation Manual—Jharkhand Police. Retrieved from: https://jhpolice.gov.in/sites/default/files/documents-reports/jhpolice_cyber_crime_investigation_manual.pdf, visited on 16 January, 2024.

4. **Case Closure:** The post-investigation stage also involves formally closing the case. This may include preparing final reports, releasing evidence to the appropriate parties, and wrapping up any administrative or legal tasks related to the investigation.[38]
5. **Presentation of Digital Evidence:** In many investigations, the findings from digital evidence must be presented clearly. This involves documenting the findings, preparing reports, and sometimes presenting evidence in court. Forensic experts need to ensure that their findings are admissible in court, as the integrity of the evidence is often challenged during legal proceedings.
6. **Archiving and Storage:** Archiving and storage in the post-investigation stage of digital forensics are critical aspects of maintaining the integrity, security, and accessibility of digital evidence after the conclusion of an investigation. Proper archiving and storage practices ensure that evidence remains unaltered and accessible for potential future reference, legal requirements, and potential appeals. Here's an in-depth look at archiving and storage post-investigation:[39]
    (a) *Data Categorization:* Categorize digital evidence based on its sensitivity, type, and relevance to the case. This categorization helps determine how each piece of evidence should be archived and stored.
    (b) *Evidence Labelling:* Every piece of digital evidence should be appropriately labelled with a unique identifier. This identifier should be used consistently throughout the investigation and post-investigation stages. Labels may include case numbers, exhibit numbers, and timestamps.
    (c) *Long-term Storage Solutions:* Invest in long-term storage solutions, such as secure servers, network-attached storage (NAS), or digital archive systems. These solutions are designed for data retention and can safeguard evidence over extended periods.
    (d) *Access Control:* Restrict access to archived evidence to authorized personnel only. Implement strong access

38 Crime Scene Investigation: A Guide for Law Enforcement, National Forensic Science Technology Center, retrieved from: https://www.nist.gov/system/files/documents/forensics/Crime-Scene-Investigation.pdf, visited on 11 January, 2024.

39 Cyber Crime Investigation Manual—Jharkhand Police. Retrieved from: https://jhpolice.gov.in/sites/default/files/documents-reports/jhpolice_cyber_crime_investigation_manual.pdf, visited on 16 January, 2024.

controls, including authentication measures, to prevent unauthorized viewing or tampering.

(e) *Retention Policies:* Develop clear retention policies outlining how long evidence should be archived and when it can be securely disposed of. Legal requirements, such as the statute of limitations, often dictate these policies.

(f) *Data Destruction:* When it's time to dispose of archived evidence, follow secure data destruction methods, such as data wiping, shredding, or physically destroying storage media.

(g) *Documentation:* Keep records of archived evidence, including its location, status, and actions taken. Documentation is vital for maintaining the chain of custody and demonstrating due diligence in evidence management.

5.5 CARDINAL RULES FOR INVESTIGATION OF DIGITAL DEVICES

The role of computer forensic professional is to collect evidence from a suspect's computer and conduct a systematic approach to determine whether or not the suspect committed a crime.[40] Before the investigator works on the case specific rules and procedures must be followed. The Cardinal rules have evolved to facilitate a forensically sound examination of computer media and enable a forensic professional to testify in the court regarding their handling a particular piece of evidence.[41]

1. **Never Mishandle the Evidence:** The first cardinal rule says to preserve the evidence, which means that the evidence should not be tampered with or contaminated. Secure evidence collection is essential to guarantee the evidential integrity and security of information. The best approach for this matter is to use a disk imaging tool. Choosing and using the right imaging tool is crucial in cyber forensics investigation.
2. **Never Work on the Original Evidence:** The second cardinal rule says not to work on the original evidence as the digital

40 Notes on Digital Forensics, retrieved from: https://www.coursehero.com/file/103323482/Digital-Forensic-Basics-Notespdf/, visited on 13 January, 2024.

41 Course on Digital Forensics, retrieved from: https://mrcet.com/downloads/digital_notes/CSE/III%20Year/12082022/DIGITAL%20FORENSICS.pdf, visited on 13 January, 2024.

evidence is very fragile. To maintain the integrity of the digital evidence and any unknowing alteration, preserve the original evidence in its pristine condition. To preserve the original evidence, a forensic copy or imaging of the original data is done using specialized software and a write blocker so that integrity of evidence is not altered. The analysis is done now on forensic copy of evidence.[42]

The original evidence is to be preserved into safe custody. It is easier to work on the original evidence and the cost related to it is also low. But, if analyzed directly, the digital evidence will lose its integrity, and authenticity and will not be admissible in any court.

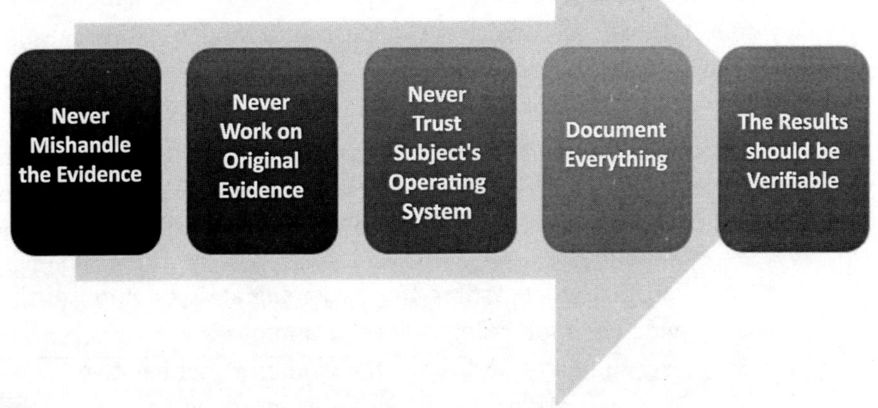

Figure 5.9 Cardinal Rules of an Investigation.

3. **Never Trust the Subject's Operating System:** Computer criminals can modify the routine operating system commands to perform destructive commands. Using the subject's operating system could quickly destroy data with just a few keystrokes. When the subject computer starts, booting to a hard disk overwrites and changes evidentiary data. To ensure that data is not altered, experts need to monitor the subject's computer

42 Pathshala, Computer Forensics, retrieved from: https://epgp.inflibnet.ac.in/epgpdata/uploads/epgp_content/law/07._information_and_communication_technology_/12._computer_forensics/et/5747_et_12_et.pdf, visited on 16 January 2024.

during initial bootstrap to identify the correct key to access the CMOs setup.

4. **Document Everything:** Chain of Custody is a documented and chronological record-keeping process that tracks the handling, possession, and transfer of physical evidence or items of potential evidentiary value from the moment they are collected until they are presented in a court of law. Chain of Custody is a documented and chronological record-keeping process that tracks the handling, possession, and transfer of physical evidence or items of potential evidentiary value from the moment they are collected until they are presented in a court of law.[43] The chain of custody is essentially a paper trail that accounts for the whereabouts and condition of evidence at every stage of its journey through the legal system. The chain of custody is essentially a paper trail that accounts for the whereabouts and condition of evidence at every stage of its journey through the legal system. To document the evidence chain-of-evidence form is created. It serves the following functions:
   (a) Identify the evidence.
   (b) A legal authority copy should be obtained.
   (c) Chain of custody including initial count of evidence to be examined,
   (d) Information regarding the packaging and condition of the evidence upon receipt by the examiner.
   (e) Lists the dates and times the evidence was handled.
   (f) Documentation should be preserved according to the examiner's agency policy.

5. **The Results should be Repeatable and Verifiable by a Third Party:** The fifth cardinal rule says that the analysis done on the evidence should be completely audited by the third party. To establish the integrity of information a cryptographic hash value, such as MD5 or SHA-1 are calculated so that it can be proven to the courts. Chain of custody forms are created if evidence is used in court or verified by any third party. The same process can be conducted and verified by any expert or person. Recognizing the fragile nature of the digital data, major task is to preserve the evidence against accidental or intentional manipulation.[44]

---

43 Chain of Custody, retrieved from: https://auth.geeksforgeeks.org/roadBlock_v2.php, visited on 18 January, 2024.

44 Mozid, M. A., & Yesmen, N. *Journal of Advanced Forensic Sciences.*

5.6 CONCLUSION

Therefore, it can be summarized that digital forensic investigation is highly relevant in today's technologically driven world, where digital devices and networks play a crucial role in various aspects of our lives.[45] This investigative process involves collecting, analyzing, and preserving electronic evidence to uncover and prevent cybercrime or other illicit activities. The relevance of digital forensic investigation can be summarized in several vital points:[46]

(i) *Legal Investigations and Evidence:* Digital evidence is becoming increasingly crucial in court cases. Digital forensic investigations give lawyers and law enforcement organizations the proof they need to prove their cases. This includes evidence of fraud, theft of intellectual property, cybercrimes, and other online offences.

(ii) *Incident Response and Recovery:* To effectively address security incidents, organizations use digital forensics. They can determine the perpetrators of a security breach, assess the extent and impact of the incident, and take the necessary remedial action by examining digital evidence. By doing this, downtime and monetary losses are reduced.

(iii) *Cybercrime Prevention and Detection:* Digital forensics helps prevent and detect cybercrimes by examining digital evidence left behind on devices and networks. Investigators can identify unauthorized access, data breaches, malware attacks, and other cyber threats to safeguard individuals, organizations, and governments from malicious activities.[47]

(iv) *National Security and Counterterrorism:* Digital forensics is critical in both these domains. Investigators track terrorist activity, gather intelligence on cyber-based threats to national security, and trace and thwart potential threats by analyzing digital evidence.

(v) *Protection of Intellectual Property:* By detecting and following instances of unapproved use or distribution, digital forensics helps to protect intellectual property. This is especially important

45 Horsman, G. (2023). The importance of digital evidence strategies. *Wiley Interdisciplinary Reviews: Forensic Science*, e1507.

46 Maras, M.-H. (2014). Computer Forensics: Cybercriminals, Laws, and Evidence. Jones & Bartlett Learning.

47 Nelson, B., Phillips, A., & Enfinger, F. (2014). Guide to Computer Forensics and Investigations. Cengage Learning.

for sectors where intellectual property is highly valued, like technology, pharmaceuticals, and entertainment.

(vi) *Collaboration between the Public and Private Sectors:* Public and private sector organizations are frequently involved in digital forensic investigations. This cooperation depends on a thorough and successful strategy to counter cyber threats and guarantee digital infrastructure security.

(vii) *Data Recovery and Reconstruction:* In cases of accidental data loss or system failures, digital forensics plays a crucial role in data recovery. Investigators use specialized tools and techniques to reconstruct lost or damaged digital information, helping individuals and organizations retrieve critical data.[48]

(viii) *Regulatory Compliance:* Many industries must adhere to specific data protection and privacy regulations. Digital forensics assists organizations in ensuring compliance with these regulations by providing the necessary tools and methodologies to investigate and report on security incidents.[49]

(ix) *Corporate Misconduct Investigations:* Employers utilize digital forensics to investigate employee misbehaviour, including data theft, illegal access, and policy violations. By carefully reviewing digital trails, employers can decide whether to take disciplinary action or pursue legal action.

In conclusion, the relevance of digital forensic investigation extends across various domains, including law enforcement, legal proceedings, corporate security, and national defence. As technology advances, the importance of digital forensics will only grow in addressing the evolving landscape of cyber threats and digital crimes.

LET'S RECALL

- The pre-investigation phase is crucial for laying the groundwork for a successful digital forensic investigation. It involves defining the scope, objectives, and assembling adequate tools and equipment,

48 Karie, N. M., Kebande, V. R., Venter, H. S., & Choo, K. K. R. (2019). On the importance of standardising the process of generating digital forensic reports. *Forensic Science International: Reports*, 1, 100008.

49 Lillis, D., Becker, B., O'Sullivan, T., & Scanlon, M. (2016). Current challenges and future research areas for digital forensic investigation. *arXiv preprint arXiv:1604.03850*.

including specialized hardware and software used in digital forensics.

- Standard operating procedures and cardinal rules are essential to maintain the integrity of the investigation, ensuring that evidence is not mishandled. Rules include:
  - not mishandling evidence,
  - not working on the original evidence,
  - not trusting the subject's operating system,
  - documenting everything, and
  - ensuring results are repeatable and verifiable by a third party.
- Digital evidence can be found in a wide range of electronic devices and systems, such as computers, mobile devices, servers, network logs, email servers, IoT devices, and more.
- Data acquisition in digital forensics includes two categories: live acquisition and static acquisition.
  - Live acquisition involves collecting data from a running device and is used for capturing volatile data.
  - Static acquisition focuses on non-volatile data stored on storage media, creating forensic images for analysis.
- Data acquisition methods like disk-to-image, disk-to-disk copy, and logical disk-to-disk are employed based on specific scenarios.
- Preservation of digital evidence, transportation, and examination are key aspects of this during the investigation phase. Preserving digital evidence is critical due to its fragile nature, and write-blocking mechanisms are employed to prevent tampering.
- Techniques used during the investigation to examine data include data recovery, keyword searching, metadata examination, timeline analysis, and data carving.
- Challenges during this phase include:
  - handling a vast data volume,
  - dealing with data encryption,
  - privacy concerns,
  - cybersecurity risks,
  - data fragmentation, and
  - keeping up with emerging technologies.
- The post-investigation stage follows the active analysis of digital evidence and includes documentation and reporting, expert testimony, legal compliance, case closure, and archiving and storage.

A. CHOOSE THE CORRECT OPTION

1. What aspect of digital evidence analysis involves keyword searching to identify specific information within data sets?
   (a) Data recovery and reconstruction
   (b) Metadata examination
   (c) Timeline analysis
   (d) Keyword searching
2. What is the purpose of data carving in digital forensics?
   (a) Collecting data from running computers
   (b) Reassembling fragmented or unallocated data
   (c) Decrypting encrypted files
   (d) Creating forensic images of storage media
3. What is the cardinal rule that emphasizes not altering digital evidence during an investigation?
   (a) Never work on the original evidence
   (b) Trust the subject's operating system
   (c) Document everything
   (d) Never mishandle the evidence
4. Which type of data acquisition method is suitable for capturing volatile data from a running computer?
   (a) Live data acquisition
   (b) Static data acquisition
   (c) Logical disk-to-disk copy
   (d) Disk-to-image file
5. Why is it important to restrict access to archived digital evidence?
   (a) To prevent any evidence from being collected
   (b) To encourage collaboration among investigators
   (c) To safeguard the integrity of the evidence
   (d) To speed up the investigation process

B. FILL IN THE BLANKS

6. The tool used to preserve the original evidence and prevent data tampering is called a ____________.
7. The fifth cardinal rule emphasizes that the results of analysis should be repeatable and verifiable by a ______________ party.

8. During the post-investigation stage, digital forensic experts must ensure that all actions taken during the investigation comply with legal standards to maintain the ___________ of evidence.
9. Forensic experts should never trust the subject's operating system because it can be modified to perform ___________ commands.
10. The first cardinal rule of digital forensics is to ___________ the evidence to preserve its integrity and security.

Answer Key

1. (d) Keyword searching
2. (b) Reassembling fragmented or unallocated data
3. (a) Never work on the original evidence
4. (a) Live data acquisition
5. (c) To safeguard the integrity of the evidence
6. Write-blocker
7. third
8. Admissibility
9. Destructive
10. preserve

C. ANSWER THE FOLLOWING

1. What are the cardinal rules of a digital forensic investigation, and how do they ensure the forensic soundness of an examination?
2. Explain the archiving and storage practices involved in the post-investigation stage, emphasizing data categorization, retention policies, and access control.
3. What is the significance of the pre-investigation phase in digital forensics, and how does proper planning and preparation enhance the chances of a successful outcome?
4. What is the role of digital evidence preservation, and how can experts ensure the data remains unchanged during the investigation process?
5. Explain the importance of a well-defined chain of custody in handling digital evidence.

D. THINK OUTSIDE THE BOX

1. How might technological advancements impact the digital forensics investigation process, and what creative solutions could you develop as a digital forensic expert to address potential challenges?

CHAPTER 6

Electronic Evidence Legal Provisions and Landmark Case Laws in India

A Glimpse into the Chapter

- Legal Provisions on Electronic Evidence
- Landmark Cases on Electronic Evidence

In today's digital age, electronic evidence plays a crucial role in legal proceedings, offering valuable insights and documentation in various cases, such as cybercrime, intellectual property disputes, and corporate fraud. Like many other countries, India has recognised electronic evidence's significance in courtrooms. However, its admissibility is subject to a specific legal framework governed by the Indian Evidence Act, 1872, and the Information Technology Act, 2000 (as amended in 2008), among other relevant statutes. This chapter explores the legal framework and implications of the admissibility of electronic evidence in India while analysing prominent judgements.

6.1 LEGAL PROVISIONS ON ELECTRONIC EVIDENCE: AN INTRODUCTION

Electronic evidence is significant in the investigation process, especially in today's technologically driven environment. It has been a boon to society and the legal system, however it does present its own set of challenges, the most significant being its admissibility. Here are the key provisions regarding the admissibility of electronic evidence in India:

Section 65A of the Indian Evidence Act 1872

It establishes specific provisions for electronic record evidence, stating that "the contents of electronic records may be proved in accordance with the provisions of Section 65B."

Section 65B of the Indian Evidence Act 1872

It goes into detail about the admissibility of e-evidence.

(i) Section 65B(1) specifies that anything held in the form of an electronic record, whether printed on paper, saved, or recorded, can be recognized as admissible evidence only if it meets the standards outlined in later Section 65B provisions.

(ii) Section 65B(2) discusses the conditions under which the computer from which an electronic record was made can be considered a reliable source. The computer must be used on a regular basis, and it must be the one that was regularly used to save the information on the activity in question. Furthermore, the information obtained from the computer must have been supplied into it in the normal course of business. Finally, the computer must have been operational during the relevant time period.

(iii) Section 65B(3) says that for the purposes of evidence, a group or combination of computers linked during the relevant time period shall be treated as a single computer. Additionally, if one or more computers act in succession across the time period in question, they are to be treated as a single computer.

(iv) Section 65B(4) specifies that when the type of evidence in a proceeding is that of electronic form, a certificate containing particulars of the device or anything dealing with Sub-section 2 of the Section must be issued. The certificate must be authorized by a person who holds a responsible official position about the device's specifics.

6.2 LANDMARK CASE LAWS ON ELECTRONIC EVIDENCE IN INDIA: A BRIEF REFLECTION

R.M. MALKANI V. STATE OF MAHARASHTRA[1]

Supreme Court of India | September 22, 1972 | 2 Judge Bench

Justice A.N. Ray, Justice I.D. Dua

Case Relevance:

Admissibility of Tape-Recorded Evidence.[2]

Statement of Facts:

1. The case longs back to 1964 when a person (Jagdish Prasad Ramnarayan Khandelwal) was admitted to a nursing home for acute appendicitis (as diagnosed by Gynecologist Dr. Adatia) which was further classified as acute appendicitis with generalized peritonitis (as diagnosed by Dr. Shantilal J. Mehta), an operation was performed, post which, he died.
2. The dead body was sent for disposal without post mortem by the appellant. The death being post operation, a request for an inquest was made from the police station wherein the Coroner's Court of Bombay made an order for all the doctors who treated and operated on the patient to explain the death.
3. On the appellant's request, Dr. Adatia was urged to consult with Dr. Motwani in order to resolve the case's technical concerns. He requested that Dr. Motwani demand ₹ 20,000 from Dr. Adatia. The appellant eventually lowered his demand to 10,000, but Dr. Adatia once more refused to provide him any illegal gratification. Further, complaint was filed with the Anti-Corruption Bureau by Dr. Motwani.
4. The director of the Anti-Corruption Branch along with Assistant Commissioner of Police, arranged to attached a "Tape Recording Equipment" to Dr. Motwani's telephone and asked him to call the appellant in their presence. All the telephonic conversations thereon were recorded.
5. The appellant was found guilty by the High Court of Bombay. A fine of ₹ 10,000 was imposed on him and six months of simple imprisonment if the fine wasn't paid. Hence the appeal.

Procedural History:

The case was previously decided by High Court of Bombay on 09 October 1969.

---

1 (1973) 1 SCC 471.

2 Before the enactment of the Information Technology Act 2000 & addition of special provisions Section 65A & 65B under the Indian Evidence Act, 1872.

1. The trial Court and the High Court unanimously stated that Dr. Motwani's and Dr. Adatia's evidence needed to be verified.
2. The appellant was charged under Sections 161, 385, and 420 of the Indian Penal Code, read with Section 511.
3. The Appellant was convicted by the High Court under Sections 161 and 385 of the Indian Penal Code. A fine of ₹ 10,000 and an additional six months of simple imprisonment were imposed by the High Court of Bombay in the event that the fine was not paid.

Issues:

1. Whether the tape-recorded conversation is admissible?
2. Whether the evidence was obtained illegally u/s 25 of the Indian Telegraph Act, 1885[3]? If an illegally obtained tape-recorded evidence is admissible?
3. Whether the tape-recorded conversation be considered as evidence when recorded in the presence of a Police Officer u/s 161[4] & 162[5] of The Code of Criminal Procedure, 1973?

---

3 **Intentionally damaging or tampering with telegraphs. If any person, intending:**
(a) to prevent or obstruct the transmission or delivery of any message, or
(b) to intercept or to acquaint himself with the contents of any message, or
(c) to commit mischief, damages, removes, tampers with or touches any battery, machinery, telegraph lines, post or other thing whatever, being part of or used in or about any telegraph or in the working thereof, he shall be punished with imprisonment for a term which may extend to three years, or with fine, or with both.

4 **Examination of witnesses by police:**
(1) Any police officer making an investigation under this Chapter, or any police officer not below such rank as the State Government may, by general or special order, prescribe in this behalf, acting on the requisition of such officer, may examine orally any person supposed to be acquainted with the facts and circumstances of the case.
(2) Such person shall be bound to answer truly all questions relating to such case put to him by such officer, other than questions the answers to which would have a tendency to expose him to a criminal charge or to a penalty or forfeiture.
(3) The police officer may reduce into writing any statement made to him in the course of an examination under this section; and if he does so, he shall make a separate and true record of the statement of each such person whose statement he records.

5 **Statements to police not to be signed: Use of statements in evidence:**
(1) No statement made by any person to a police officer in the course of an investigation under this Chapter, shall, if reduced to writing, be signed by the person making it; nor shall any such statement or any record thereof, whether in a police diary or otherwise, or any part of such statement or record, be used for any purpose, save as hereinafter.
(2) Nothing in this section shall be deemed to apply to any statement falling within the provisions of Clause(1) of Section 32 of the Indian Evidence Act, 1872 (1 of 1872), or to affect the provisions of Section 27 of that Act.

4. Whether Tape recorded conversation infringes Article 20(3)[6] & Article 21[7] of the Constitution of India?

Rule of Law/Legal Principles/Provisions Applied or Relevant to the case:

1. Section 25 of The Indian Telegraph Act, 1885.
2. Section 161 & 162 of The Code of Criminal Procedure, 1973.
3. Section 7, 8, 146 &153 of The Indian Evidence Act, 1872.
4. Article 20(3) & 21 of The Constitution of India.

Prominent Arguments by the Advocates:

1. **Appellant**
   (a) The counsel for appellant challenged the admissibility of tape-recorded conversations.
   (b) The said evidence infringes Articles 20(3) & 21 of the Constitution of India.
   (c) Attaching tape recording instrument to the telephone instrument is an offence u/s 25 of The Indian Telegraph Act, 1885.

Judgement—Addressing the Issues:

1. *Whether the tape-recorded conversation is admissible?*

- It was stated that tape-recorded conversation is admissible in the court of law, provided:
  - It is relevant to the matter.
  - The voice is identified.
  - The authenticity of tape-recorded conversation is proved.
- "A contemporaneous tape record of a relevant conversation is a relevant fact and is admissible under Section of 8 The Evidence Act. It is res gestae. It is also comparable to a photograph of a relevant incident. The tape-recorded conversation is therefore a relevant fact and is admissible under Section 7 of the Evidence Act."

2. *Whether the evidence was obtained illegally u/s 25 of the Indian Telegraph Act, 1885? If an illegally obtained tape-recorded evidence is admissible?*

- It was stated that there is no violation of the above-stated provision in the present case.

6 No person accused of any offence shall be compelled to be a witness against himself.

7 Protection of life and personal liberty. No person shall be deprived of his life or personal liberty except according to procedure established by law.

- It was held that as long as the evidence's relevance and authenticity are established, it will be admissible in court even if it was obtained illegally.
- Even if the individual whose conversation is being recorded is unaware that the tape recording device is attached, if the conversation is voluntary and there is no coercion, the recorded conversation is admissible.

3. *Whether the tape-recorded conversation be considered as evidence when recorded in the presence of a Police Officer u/s 161 & 162 of The Code of Criminal Procedure, 1973?*
   - It was held that in the current case, the police officer, with Dr. Motwani's approval, attached the tape-recording device to the telephone device.
   - Instead of directly hearing the oral interaction between Dr. Motwani and the appellant, the Police Officer taped the conversation with a tape recorder.
   - Furthermore, in the present case, the appellant was not compelled to confess or even speak at the time of conversation.

4. *Whether Tape recorded conversation infringes Article 20(3) & Article 21 of the Constitution of India?*
   - It was held that the telephonic conversation of an innocent citizen will be protected against illegal tapping or recording, but the said protection is not for guilty citizen.

Final Decision:

1. The appeal was dismissed.
2. The appellant was ordered to serve his sentence.

ZIYAUDDIN BURHANUDDIN BUKHARI V. BRIJMOHAN RAMDASS MEHRA & ORS.[8]

Supreme Court of India | April 25, 1975 | 3 Judge Bench

Justice M.H. Beg, Justice A. Alagiriswami, Justice N. L. Untwalia

Case Relevance:

Admissibility of tape-recordings of election speeches.[9]

---

8 (1976) 2 SCC 17

9 Before the enactment of the Information Technology Act 2000 & addition of special provisions Section 65A & 65B under the Indian Evidence Act, 1872.

Statement of Facts:

1. The case is associated with Maharashtra State Assembly from Kumbharwada constituency held in the year 1972.
2. It was claimed that throughout his election campaign, he delivered speeches designed to make voters believe that if they support the opposition candidate, they will face divine wrath or spiritual judgment.
3. The voter alleged that the appellant engaged in corrupt behavior as described in Section 123, Sub-sections S(2), (3), and (3A) of the Act throughout the course of his election.
4. To prove the same, tape recordings of the election speeches were submitted.

Key Issue(s):

Whether the tape-recording of election speech is admissible?

Rule of Law/Legal Principles/Provisions Applied or Relevant to the case:

1. Section 3 of The Indian Evidence Act, 1872.
2. Section 6 of The Indian Evidence Act, 1872.
3. Section 7 of The Indian Evidence Act, 1872.
4. Section 159 of The Indian Evidence Act, 1872.
5. Section 160 of The Indian Evidence Act, 1872.

Judgement—Addressing the Issues:

1. It was held that *"Taperecords of speeches are "documents" under Section 3 of the Evidence Act and stand on no different footing than photographs."*
2. When a person's speech or statement is in issue, there could be no more direct or better evidence of it than its taperecord, if its authenticity is duly established.
3. Taperecords are admissible in evidence on satisfying the following conditions:
   (a) The voice of the person alleged to be speaking must be duly identified by the maker of the record or by others who know it.
   (b) Accuracy of what was actually recorded has to be proved by the maker of the record and satisfactory evidence, direct or circumstantial, has to be there so as to rule out possibilities of tampering with the record.
   (c) The subject-matter recorded has to be shown to be relevant according to rules of relevancy found in the Evidence Act.

Final Decision:

1. The appeal was dismissed.

TUKARAM S. DIGHOLE V. MANIKRAO SHIVAJI KOKATE[10]
Supreme Court of India | February 5, 2010 | 2 Judge Bench
Justice D.K. Jain, Justice P. Sathasivam

Case Relevance:
Admissibility of VHS Cassette

Statement of Facts:

1. The parties contested for election to Lok Sabha in the year 2004 wherein the respondent succeeded and was declared elected.
2. The appellant was not satisfied with the results and thus filed an election petition challenging the respondent's election on various grounds stated under the Representation of People Act, 1951. (Primarily on the ground of corrupt practices and communal votes)
3. A VHS Cassette, allegedly obtained from the Election Commission of India containing true reproduction of speeches delivered by the respondent was placed on record by the appellant.
4. The said petition was dismissed by the High Court of Bombay, hence the appeal.

Procedural History:
The case was previously decided by High Court of Bombay on 25 January 2008[11].

1. The hon'ble High Court held "*There is no evidence to indicate that the VHS cassette was obtained from the election commission. The petitioner has failed to establish that the VHS cassette was a public document since they did not provide evidence supporting the election commission's office receipt. The Indian Evidence Act does not require that a public document be proven. However, it must be brought on record as evidence.*"

Issues:
Whether mere production of an audio cassette (assuming it to be a certified copy) is conclusive of the fact that its content are true and correct?

10 (2010) 4 SCC 329
11 2008 SCC OnLine Bom 79

Rule of Law/Legal Principles/Provisions Applied or Relevant to the case:

1. Section 74 of The Indian Evidence Act, 1872.
2. Section 76 of The Indian Evidence Act, 1872.

Prominent Arguments by the Advocates:

1. **Appellant**
   (a) The counsel for appellant argued that because the VHS Cassette was a public record and the respondent had not expressly denied its contents, no additional proof was needed to establish the cassette's validity.
   (b) It was submitted in the opinion of the learned counsel, that the High court's decision to reject the evidence presented by the appellant in the form of the aforementioned cassette constituted a major legal error.
2. **Respondent**
   (a) It was argued that, in addition to the lack of a specific pleading in the election petition regarding the method of obtaining the cassette in question, the appellant was required to provide convincing evidence that the speeches captured on the cassette were delivered by the respondent and his agents in order to trigger the provisions of Section 123 of the Act. Thus, supporting the High Court's decision.

Judgement:

1. It was held that the petitioner has not shown any evidence that the VHS cassette was a genuine reproduction of the original speeches.
2. The hon'ble bench stated that "*There is also no doubt that the new techniques and devices are the order of the day. Technology based on audio and videotapes has grown to be a potent medium that might serve as a significant piece of evidence. At the same time, tapes and cassettes are more prone to tampering and alteration, therefore such evidence needs to be treated with caution.*"

Final Decision:

1. The appeal was dismissed.

ANVAR P.V. V. P.K. BASHEER & ORS.[12]

Supreme Court of India | September 18, 2014 | 3 Judge Bench

Chief Justice, R.M. Lodha, Justice Kurian Joseph, Justice Rohinton Fali Nariman

Case Relevance:

Interpretation of Section 65B—Admissibility of Electronic Evidence

Statement of Facts:

1. This case is associated with the Kerela Legislative Assembly general election that took place in April 2011.
2. The respondent was elected in this election.
3. The appellant, who contested independently and was backed by the Left, received the second majority of votes.
4. The appellant filed an election petition to nullify the respondent's election on the grounds of corrupt practices under Sections 100(1)(b), 123(2)(ii), and 123(4) of the Representation of Peoples Act.
5. Recording of election campaign speeches were placed as evidence.
6. The election petition was dismissed by the Kerala High Court.

Procedural History:

The case was previously decided by High Court of Bombay on 13 April 2012[13].

1. The High Court held that election of the first respondent cannot be declared void or set aside.
2. The Election Petition was dismissed with cost of ₹ 2,000/- to the first respondent.

Issues:

1. Under what conditions can electronic record be admitted as primary and secondary evidence?
2. What are the pre-requirements to produce secondary evidence of electronic records? Is it necessary to produce certificate u/s 65B(4)?
3. Whether an electronic record be admitted as secondary evidence by the way of oral evidence?

---

12 (2014) 10 SCC 473

13 2012 SCC OnLine Ker 31838

Rule of Law/Legal Principles/Provisions Applied or Relevant to the case:

1. Section 22A of the Indian Evidence Act, 1872.
2. Section 45A of the Indian Evidence Act, 1872.
3. Section 59 of the Indian Evidence Act, 1872.
4. Section 62 of the Indian Evidence Act, 1872.
5. Section 65A of the Indian Evidence Act, 1872.
6. Section 65B of the Indian Evidence Act, 1872.

Judgement—Addressing the Issues:

1. The bench stated that *"Genuineness, veracity or reliability of the evidence is seen by the court only after the stage of relevancy and admissibility. These are some of the first principles of evidence."*
2. Under what conditions can electronic record be admitted as primary and secondary evidence?
   (a) If primary evidence of electronic record is adduced, it is admissible u/s 62, without compliance with conditions u/s 65B.
   (b) If secondary evidence of electronic record is adduced, it shall satisfy the conditions prescribed u/s 65B.
3. What are the pre-requirements to produce secondary evidence of electronic records? Is it necessary to produce certificate u/s 65B(4)?
   (a) It was held that if the electronic record being produced is a copy of statement and not the original electronic record, i.e. in a CD, VCD, Pen Drive, etc. it must be accompanied by certificate prescribed u/s 65B(4).
   (b) Secondary Electronic record cannot be admitted as evidence in absence of the said certificate.
4. Whether an electronic record be admitted as secondary evidence by the way of oral evidence?
   (a) It was held that electronic record be admitted as secondary evidence by the way of oral evidence if the requirements u/s 65B are not complied with.
5. Sections 65A & 65B are a code in itself with respect to Electronic Records. The requirement to produce certificate is necessary.
6. Sections 63 & 65 are general laws whereas Section 65A & 65B are special laws and as per rule of law, special law shall prevail general law.
7. The requirement of Section 65B are intended to safeguard source and authenticity of electronic records (as they are more susceptible to tampering, alteration, etc.).

Final Decision:

1. The appeal was dismissed.
2. The election of the first respondent cannot be declared void or set aside.

Additional Information:

1. The judgement was upheld by *Arjun Panditrao Khotkar v. Kailash Kushanrao Gorantyal.*

TOMASO BRUNO V. STATE OF UP[14]

Supreme Court of India | January 20, 2015 | 3 Judge Bench

Justice Anil R. Dave, Justice Kurian Joseph, Justice R. Banumathi

Case Relevance:

Rule of Best Evidence in the age of Science & Technology.

Statement of Facts:

1. The case begins in December 2009 when three tourists of Italian nations visit India from London.
2. In February 2019, they reach Varanasi where one of them got ill and was later declared 'brought dead' when taken to the nearest hospital.
3. A complaint regarding the death was filed.
4. Upon investigation the two tourists were accused and arrested.
5. The trial court and High Court convicted and sentenced the accused.

Procedural History:

The case was previously decided by High Court of Allahabad on 04 October 2012[15] and by Trial Court on 23 July 2011[16].

1. The trial court convicted and sentenced the accused under Section 302 read with 34 of the Indian Penal Code.
2. The accused were sentenced to life imprisonment along with ₹ 25,000 fine each.
3. The High Court dismissed the appeal and affirmed the judgment and order of the trial court.

---

14 (2015) 7 SCC 178

15 2012 SCC OnLine All 4075

16 23rd July, 2011 passed by the Additional Sessions Judge, Court No. 13, Varanasi in Session Trial No. 299 of 2010 (*State of UP* v. *Tomaso Bruno*) arising out of case Crime No. 34 of 2010.

Issues of Interest:

1. Whether the production of CCTV footage is necessary when such is available?

Rule of Law/Legal Principles/Provisions Applied or Relevant to the Case:

1. Section 65 of Indian Evidence Act, 1872.
2. Section 65A of Indian Evidence Act, 1872.
3. Section 65B of Indian Evidence Act, 1872.
4. Section 114 of Indian Evidence Act, 1872.

Prominent Arguments by the Advocates:

1. **Appellant**
   (a) It was argued that as CCTV footage is a crucial piece of evidence, its non-production seriously undermines the prosecution's case and renders it impossible to proceed with the case.
   (b) Further, it was argued that the lower courts should have picked up on the flawed investigation, lack of CCTV footage, sim details, and other oversights.
3. **Respondent**
   (a) It was argued that the medical evidence indisputably implied the guilt of the accused people, and that after taking into account the circumstances and the evidence presented by the prosecution, the lower courts appropriately convicted the appellants.
   (b) Accordingly, the concurrent conclusions made by the lower courts were said to be unaffected by any additional evidence.

Judgement:

1. The hon'ble bench stated that *"With the advancement of information technology, scientific temper in the individual and at the institutional level is to pervade the methods of investigation."*
2. The use of electronic evidence in court is now necessary to prove an accused person's guilt or the defendant's liability due to the growing influence of technology in daily life.
3. Electronic documents are accepted as material evidence in the strictest sense.
4. Since CCTV cameras were installed in the prominent places, CCTV footage would have been best and a strong piece of evidence.

Final Decision:

1. The appeal was allowed.
2. The conviction was set aside and appellants were released.

Additional Information:

1. As per the Arjun Panditrao Judgement, the court's view of admissibility of CCTV footage as secondary evidence u/s 65 of the act was incorrect statement of law followed by citing Navjot Sandhu judgment that was overruled by the Anvar PV judgement (also not referred).
2. The decision was overruled by *Arjun Panditrao Khotkar v. Kailash Kushanrao Gorantyal.*

SONU ALIAS AMAR V. STATE OF HARYANA[17]

Supreme Court of India | July 18, 2017 | 2 Judge Bench

Justice S.A. Bobde, Justice L. Nageswara Rao

Case Relevance:

Objection regarding mode/method of proof of electronic record at appellate stage.[18]

Statement of Facts:

1. In December 2005, a man was reported missing. Thereafter, in January 2006, a call for ransom of 1 crore was received.
2. Upon investigation, the accused were located and arrested and their mobile phones were confiscated.
3. The accused suffered disclosure of statements during the course of investigation and revealed that they abducted the victim and later on buried his body.
4. Thereafter, CDRs (Call Detail Records) of the seized mobile phones were collected from the nodal officers of mobile companies.
5. Both the Trial Court and the High Court convicted and sentenced the accused.

Procedural History:

The case was previously decided by High Court of Punjab & Haryana on 11 October 2012[19].

---

17 (2017) 8 SCC 570

18 Such objection was raised for the first time before the hon'ble Supreme Court.

19 *Surender v. State of Haryana*, CRA-D No. 1061-DB of 2010. Order dated 11-10-2012 (P&H).

1. The trial court convicted and sentenced the accused for life imprisonment.
2. The High Court confirmed the trial court's decision. The disclosure statements, subsequent recoveries, and the CDRs of the mobile phones were relied upon by the High Court to support the trial court's conclusions and they were also regarded as reliable sources of evidence.

Issues:

1. Whether the objection regarding mode/method of proof of electronic record be raised at appellate stage?

Rule of Law/Legal Principles/Provisions Applied or Relevant to the Case:

1. Section 5, The Indian Evidence Act, 1872.
2. Sections 59 to 65 B, The Indian Evidence Act, 1872.

Prominent Arguments by the Advocates:

1. **Appellant**
   (a) It was contended that the CDRs are not admissible u/s 65B of the Act because they did not adhere to the requirement of a certificate u/s 65B(4).
   (b) It was argued that the argument he made had to do with the document's admissibility rather than the method of proof.
   (c) Furthermore, it was submitted that no objection was raised when the Trial Court heard the CDRs as evidence. According to the record, no such objection appears to have been raised even at the High Court's appellate stage.
2. **Respondent**
   (a) It was contended that the CDRs were introduced into evidence without any objection from the defense.
   (b) He argued that the accused cannot bring up the CDRs' admissibility at the appellate stage.
   (c) It was submitted that there are two types of objections to the admissibility of documents: first, that a document is inadmissible as evidence by itself, and second, where the issue is to the manner or method of the document's proof. Since it cannot be said that CDRs are per se inadmissible in evidence, it was contended that the current case points towards second class.

Judgement:

1. The Hon'ble Supreme Court held that *it is nobody's case that CDRs which are a form of electronic record are not inherently admissible in evidence*. The argument is that they were marked before the Trial Court without a certificate, as required by Section 65B. (4). An objection to the methodology or technique of proof must be raised at the time the document is marked as an exhibit, not later.

Final Decision:

1. The decision of trial court and the High Court were upheld.
2. The appeal was dismissed.

SHAFHI MOHAMMAD V. STATE OF HIMACHAL PRADESH[20]

Supreme Court of India | April 03, 2018 | 2 Judge Bench

Justice Adarsh Kumar Goel, Justice Rohinton Fali Nariman

Case Relevance:

Validity of use of Videography at a scene of crime, Applicability of Section 65B(4).

Statement of Facts:

1. The case longs back to 2008, when Shafhi Mohammad (Accused) was found in possession of drugs (charas). He was accused u/s 20 of NDPS Act, 1885.
2. On 16.12.2008, Inspector, SHO PS Chamba, Himachal Pradesh stopped a vehicle that was driven by Rakesh Kumar (accused) and had Shafhi Mohammad (accused) on the passenger seat. Shafhi Mohammad carried a bag on his lap. The bag found in possession of the accused had charas in it.
3. The Trial court convicted Shafhi Mohammad.
4. An appeal was filed in the High Court of Himachal Pradesh wherein the trial court's decision was affirmed.
5. Although, during the course of consideration of the matter, question on "*whether videography of the scene of crime or scene of recovery during investigation should be necessary to inspire confidence in the evidence collected*" arose.
6. A Petition was filed in the Hon'ble Supreme Court of India seeking permission for use of videography of crime scene.

20 (2018) 2 SCC 801

Procedural History:

1. The prosecution proved the involvement of accused in the crime beyond any reasonable doubt. Thus, trial court convicted and sentenced the accused u/s 20 and 29 of NDPS Act, 1885 vide judgement dated 15.10.2009.
2. The appeal was then decided by the High Court of Himachal Pradesh[21] on 26.06.2014 wherein the decision of the trial court was affirmed. The High Court stated that *"The prosecution has proved beyond any reasonable doubt that the accused were involved in the crime. Statements of the police officials are consistent, credible and inspire confidence. They had no animosity with the accused. We find no reason to interfere with the well-reasoned judgment rendered by the learned trial court."*

Issues:

1. Whether use of videography be allowed at scene of crime?
2. Whether videography of scene of crime or scene of recovery during investigation necessary inspiring confidence in evidence collected?
3. Whether Section 65B(4) of Evidence Act be applicable or not?

Rule of Law/Legal Principles/Provisions Applied:

1. Section 65A of Indian Evidence Act 1872
   Special provisions as to evidence relating to electronic record.
2. Section 65B of Indian Evidence Act 1872
   States Provisions relating to Admissibility of Electronic Records.

Judgement—Addressing the Issues:

1. Sections 65A and 65B are just clarifying and procedural in character, and cannot be considered a comprehensive regulation on the issue.
2. Regardless of whether or not the standards of Section 65(B) are met, there is no restriction to introducing secondary evidence u/s 63 & 65 of the act.
3. Even if the standards under Section 65B(4) were not met, evidence could be produced under Sections 63 & 65 of the Indian Evidence Act, 1872.
4. Procedural requirement of Section 65B(4) can be relaxed and exempted for the sake of justice provided a party is not in the possession of the device.

21 *Shafhi Mohammad* v. *State of H.P.* 2014 SCC OnLine HP 5929.

5. The Centrally Driven Plan of Action developed by the Committee of Experts, as well as the schedule, were approved.
6. The Ministry of Home Affairs will establish a Central Oversight Body (COB) to oversee the continued planning and implementation of the use of videography in investigations.
7. The Centre may commence funding for the initiative, despite the fact that law and order is a state concern.
8. CCTV cameras should be deployed in all police stations and prisons to monitor human rights violations.

Additional Information:
The judgement was overruled by *Arjun Panditrao Khotkar v. Kailash Kushanrao Gorantyal*[22].

ARJUN PANDITRAO KHOTKAR V. KAILASH KUSHANRAO GORANTYAL[23]

Supreme Court of India | July 14, 2020 | 3 Judge Bench

Justice Rohinton Fali Nariman, Justice S. Ravindra Bhat, Justice V. Ramasubramanian

Case Relevance:
Admissibility of VCD (E-Evidence), Requirement of Certificate u/s 65B(4)[24], Interpretation of Section 65B.

Statement of Facts:

1. In the year 2014, two election petitions were filed by the present respondents before the Bombay High Court. It was contented that the nomination form of Returned Candidate (RC) Arjun

22 (2020) 7 SCC 1

23 (2020) 7 SCC 1

24 Section 65(4) of the Indian Evidence Act, 1872 states that—In any proceedings where it is desired to give a statement in evidence by virtue of this section, a certificate doing any of the following things, that is to say:—(a) identifying the electronic record containing the statement and describing the manner in which it was produced; (b) giving such particulars of any device involved in the production of that electronic record as may be appropriate for the purpose of showing that the electronic record was produced by a computer; (c) dealing with any of the matters to which the conditions mentioned in Sub-section (2) relate, and purporting to be signed by a person occupying a responsible official position in relation to the operation of the relevant device or the management of the relevant activities (whichever is appropriate) shall be evidence of any matter stated in the certificate; and for the purposes of this Sub-section it shall be sufficient for a matter to be stated to the best of the knowledge and belief of the person stating it.

Panditrao suffered from defects. The Nomination was filed at 3:53 PM that is after the stipulated time of 3:00 PM.

2. To prove the same, CCTV recording of the RO office were presented in the court.
3. Considering the same, in 2016, Bombay High Court by its order dated 16.03.2016, ordered the Election Commission and the concerned officers to produce the records *"including the original video recording"*, and a specific order was made which said that *"the said recordings need to be produced along with necessary certificates."*
4. Previously, the High Court of Bombay decided the matter based on substantial compliance and declared the nomination of RC Arjun Panditrao void in the impugned judgment.
5. In 2018, an appeal was filed in the Supreme Court of India for Interpreting Section 65B of the Indian Evidence Act 1972.

Procedural History:

The case was previously decided by High Court of Bombay on 24 November 2017.[25]

1. The High Court stated that the certificate required u/s 65B(4) was not produced by the officials of election commission, but the video was available on VCD.
2. The Returning Officer of the Election Commission herself came to the stands and confirmed that the video is true. Based on this, the High Court held that *"This is something more than the contents of a certificate u/s 65B(4) also because it is clarified by the responsible official person. So it is not barred by the contents of Section 65B."*
3. Based on the "Substantial Compliance" of the requirement of giving a certificate under Section 65B, the High Court held that the CD or VCDs are admissible and thus the nomination was rejected.

Issues:

1. Is it necessary for a party relying on Secondary Electronic Evidence to produce the Certificate required by Section 65B(4)?
2. At what point in the process is the Certificate envisioned by Section 65B(4) to be produced?
3. What are the Court's powers if a Certificate is not produced?

25 *Kailash Kishanrao Gorantyal* v. *Arjun Panditrao Khotkar* 2017 SCC OnLine Bom 9168.

Rule of Law/Legal Principles/Provisions Applied:

1. Section 22A of Indian Evidence Act, 1872.
   When oral admission as to contents of electronic records are relevant.
2. Sections 62 & 63 of Indian Evidence Act, 1872
   Primary & Secondary Evidence
3. Section 65A of Indian Evidence Act, 1872.
   Special provisions as to evidence relating to electronic record.
4. Section 65B of Indian Evidence Act, 1872.
   Admissibility of Electronic Records.

Prominent Arguments by the Advocates:

1. **Appellant**
   (a) The CDs/VCDs could not be admitted as evidence without a certificate u/s 65B(4).
   (b) The Tomaso Bruno judgement[26] did not notice Section 65B or Anvar P.V. judgment.[27]
   (c) Shafhi Md. Judgement[28] is contrary to larger bench of Anvar P.V. judgement and thus cannot be laid down as a good law.
   (d) The theory of "Substantial Compliance" with the provisions of Section 65B(4) could not be sustained in law because the required certificate was not issued.
2. **Respondent**
   (a) Even though the HC ordered, the certificate could not be issued by the election commission. Though several attempts were made, the certificate could not be furnished.
   (b) Section 65B is a procedural provision and absence of a certificate should not result in denial if crucial evidence.
   (c) The decision in the Shafhi Mohammad Case[29] was supported by the counsel.
   (d) Anvar P.V.[30] is a good law but intervention of judges is necessary in cases where certificate is difficult or impossible to produce.

---

26 *Tomaso Bruno and Anr.* v. *State of Uttar Pradesh* (2015) 7 SCC 178.
27 *Anvar P.V.* v. *P.K. Basheer & Ors.* (2014) 10 SCC 473.
28 *Shafhi Mohammad* v. *State of Himachal Pradesh* (2018) 2 SCC 801.
29 *Shafhi Mohammad* v. *State of Himachal Pradesh* (2018) 2 SCC 801.
30 *Anvar P.V.* v. *P.K. Basheer & Ors.* (2014) 10 SCC 473.

Judgement—Addressing the Issues:

1. Is it necessary for a party relying on Secondary Electronic Evidence to produce the Certificate required by Section 65 B(4)?
   (a) Section 65A & 65B form a comprehensive code governing the admissibility of electronic records as evidence.
   (b) Yes, in cases of secondary electronic evidence certificate is necessary.
   (c) It was further stated the it is not necessary in cases of Primary Evidence or when there is an Oral Testimony available.
2. At what point in the process is the Certificate envisioned by Section 65B(4) to be produced?
   (a) The Certificate can be produced anytime during the proceeding or as directed by the court.
   (b) Unless it does not result in serious or irreversible injury to the accused (in cases of criminal trial).
3. What are the Court's powers if a Certificate is not produced?
   (a) In the event that a certificate cannot be produced, a judge has sufficient powers and jurisdiction to order production of any document (Section 165–Evidence Act).
   (b) Also, an application can always be made to a Judge for production of such a certificate from the requisite person under Section 65B(4) in cases in which such person refuses to give it.
4. Supreme Court has issued guidelines for the internet service providers and the cell phone companies regarding maintenance of CD records and other records that maybe relevant for the purpose of seizure during investigation.
5. SC also stated that appropriate rules should be framed to for retaining data to be used in the trial of offences w.r.t. Section 67C of the IT Act which states provisions related to Preservation and retention of information by intermediaries.

Final Decision:

1. The Bombay HC judgement was Upheld.
2. The SC affirmed the decision by HC on the nomination of Arjun Panditrao and it stands rejected.

Additional Information:

1. This decision Upheld the *Anvar P.V.* v. *P.K. Basheer & Ors.*[31]
2. The judgement Overruled the *Tomaso Bruno and Anr.* v. *State of Uttar Pradesh*[32] and *Shafhi Mohammad* v. *State of Himachal Pradesh.*[33]

RAVINDER SINGH ALIAS KAKU V. STATE OF PUNJAB[34]

Supreme Court of India | May 04, 2022 | 2 Judge Bench

Justice Uday U. Lalit, Justice Vineet Saran

Case Relevance:

Admissibility of Electronic Records.

Statement of Facts:

1. In September 2009, two children were reported missing. The children were found deceased the next day. Thus, this being a case of kidnapping and murder.
2. A probe produced three suspects. During the search and seizure, the accused's cell phones and SIM cards were confiscated.
3. Call records from the aforementioned person were among the evidence that was recorded.
4. The trial court convicted the accused based on the facts, circumstances, evidence, and witnesses presented.
5. An appeal was filed, and the HC set aside the current appellant's death sentence.

Procedural History:

The case was previously decided by High Court of Punjab & Haryana on 22 February 2011.[35]

1. The trial court convicted the accused and sentenced them to death u/s 302 IPC read with 120 B IPC.
2. The Hon'ble High Court set aside the death penalty of the present appellant and sentenced him for rigorous imprisonment of 20 years u/s 302 IPC. The other two accused were acquitted.

---

31 (2014) 10 SCC 473

32 (2015) 7 SCC 178

33 (2018) 2 SCC 801

34 (2022) 7 SCC 581

35 *State of Punjab* v. *Anita* (2011 SCC OnLine P&H 17671.

Issues:

1. Whether the call records produced by the prosecution would be admissible under Sections 65A and 65B of the Evidence Act, given the fact that the requirement of certification of electronic evidence has not been complied with?

Rule of Law/Legal Principles/Provisions Applied or Relevant to the case:

1. Section 65 A of The Indian Evidence Act, 1872.
2. Section 65 B of The Indian Evidence Act, 1872.

Prominent Arguments by the Advocates:

1. **Appellant**
   (a) It was argued that the accused's conviction could not be upheld only on the basis that the prosecution demonstrated a motive through the call records.
   (b) It was contented that conviction is not possible merely based on call records, prosecution must establish the motive beyond reasonable doubt.

Judgement:

1. It was stated that when a conviction is based solely on circumstantial evidence, such evidence, and the chain of circumstances must be conclusive to sustain a conviction.
2. It was held that electronic evidence produced before the High Court, should have been in accordance with the Evidence Act and should have complied with certification requirements, for it to be admissible in the court of law.
3. Section 65 B (4) is a mandatory requirement of law and that oral evidence cannot suffice in place of such certificate.

Final Decision:

1. The appeal was allowed.
2. Conviction of the present appellant was set aside.
3. The remaining two accused's conviction was upheld.

Additional Information:

1. The *Anvar P.V.* v. *P.K. Basheer & Ors.*[36] ruling is affirmed.
2. The *Tomaso Bruno and Anr.* v. *State of Uttar Pradesh*[37] is held, *per incuriam.*

---

36 (2014) 10 SCC 473

37 (2015) 7 SCC 178

3. The *Shafhi Md.* v. *State of Himachal Pradesh*[38] is overruled.
4. The *Arjun Panditrao Khotkar* v. *Kailash Kushanrao Gorantyal*[39] ruling is affirmed.

STATE OF KARNATAKA V. T. NASEER AND ORS.[40]

Supreme Court of India | Nov 06, 2023 | 2 Judge Bench

Justice Vikram Nath, Justice Rajesh Bindal

Case Relevance:

Admissibility of Electronic Evidence in Criminal Cases.

Statement of Facts:

1. The case is rooted in a series of bomb blasts that occurred in Bangalore on July 25, 2008, leading to the loss of life and numerous injuries.
2. Various First Information Reports (FIRs) were lodged, and the accused faced multiple charges under sections of the Indian Penal Code and other relevant legislation.
3. During the investigation, electronic devices were seized upon the direction of one of the accused individuals.
4. A report was prepared by the Central Forensic Science Laboratory (CFSL) based on the data retrieved from these devices.
5. The original electronic devices were submitted to the Trial Court.

Procedural History:

The case was previously decided by High Court of Karnataka at Bengaluru on 27 January 2022.[41]

1. The Trial Court refused to admit the CFSL report as evidence due to the absence of a certificate under Section 65B of the Indian Evidence Act.
2. Following the Trial Court's rejection of the CFSL report on the grounds of lacking the Section 65B certificate, the prosecution made an application under Section 311 of the Code of Criminal Procedure to recall a witness and introduce the necessary certificate.

38 (2018) 2 SCC 801
39 (2020) 7 SCC 1
40 MANU/SC/1217/2023
41 *State of Karnataka* v. *T. Naseer and Ors.* (MANU/KA/1281/2022).

3. The Trial Court turned down this application.
4. This decision was subsequently upheld by the High Court.

Issues:

1. Whether the delay in producing the Section 65B certificate in the context of a criminal trial was reasonable and justifiable?
2. Whether the certificate mandated by Section 65B of the Indian Evidence Act is a prerequisite for admitting electronic evidence?
3. Whether the prosecution can introduce the Section 65B certificate at any stage of the legal proceedings?

Rule of Law/Legal Principles/Provisions Applied or Relevant to the Case:

1. Section 65 A of The Indian Evidence Act, 1872.
2. Section 65 B of The Indian Evidence Act, 1872.
3. Section 311 of The Code of Criminal Procedure.

Prominent Arguments by the Advocates:

1. **Appellant**
   (a) It was contended that the certificate was imperative, given the objection raised by the defense.
   (b) It was argued that the delay in producing the certificate was minimal, and it was crucial for ensuring a fair trial.
2. **Defendant**
   (a) It was argued that the delay in producing the certificate was unreasonable and would prejudice their right to a fair trial.
   (b) It was maintained that electronic evidence without the certificate should not be admitted.

Judgement:

1. The court dismissed the delay argument, asserting that the production of the Section 65B certificate did not constitute a substantial delay. It was filed promptly after the defense raised an objection and while the trial was still ongoing.
2. The court clarified the legal requirements regarding Section 65B certificates. It emphasized that these certificates are essential when electronic evidence is presented as secondary evidence but are not obligatory when the electronic record is employed as primary evidence.
3. The court stressed that the Section 65B certificate can be introduced at any stage of the trial, as long as the legal proceedings are ongoing. This ensures that the defense maintains the opportunity to challenge the evidence presented.

Final Decision:

1. The appeal was allowed.
2. Rulings of the lower court were invalidated.
3. This decision upheld the admissibility of electronic evidence in the trial, given that the requirements of Section 65B were met and that no undue prejudice would be caused to the Accused.

Additional Information:

1. The *Anvar P.V.* v. *P.K. Basheer & Ors.*[42] ruling is affirmed.
2. The *Sonu* v. *State of Haryana*[43] ruling is affirmed.
3. The *Arjun Panditrao Khotkar* v. *Kailash Kushanrao Gorantyal*[44] ruling is affirmed.

STATE (GNCT OF DELHI) V. NETRAPAL SINGH & ORS[45]

High Court of Delhi | January 09, 2024

Hon'ble Mr. Justice Amit Sharma

Case Relevance:

Admissibility of Video Cassette

Statement of Facts:

1. A writ petition[46] was filed alleging police personnel and government officials accepting bribes from bootleggers.
2. The petitioner recorded/videographed several police officials receiving bribes and submitted the video cassette to the Court. The Division bench of the court then directed the Commissioner of Police to conduct a vigilance enquiry, resulting in the registration of the present FIR.
3. The video cassette containing the recordings was sent to the Central Forensic Sciences Laboratory in Chandigarh for analysis.
4. The CFSL analysis concluded that there was no intentional tampering with the video cassette, but it identified discontinuity in the recording and blank frames.
5. Following investigation, charges were filed against nine police officials for offenses under Sections 7/13(1)(d)/13(2) of the Prevention of Corruption Act.

42 (2014) 10 SCC 473
43 (2017) 8 SCC 570
44 (2020) 7 SCC 1
45 MANU/DE/0135/2024
46 W.P.(CRL) 1897/2005

Procedural History:

The case was previously decided by learned Special Judge, Tis Hazari Courts, Delhi on 12 August 2015.[47]

1. The trial court acquitted the respondents of charges under Sections 7 and 13(1)(d) of the Prevention of Corruption Act, 1988, citing lack of proof of an actual demand for a bribe.
2. The court held that the Electronic Evidence is not admissible because of the discrepancies in the recording, suggesting potential tampering, and lack of certificate under Section 65B of the Indian Evidence Act.

Issues:

1. Whether the video recording produced be admissible, considering concerns over its originality, potential tampering, and lack of the Certificate as under Section 65B of the act?

Rule of Law/ Legal Principles/ Provisions Applied or Relevant to the Case:

Section 65 A of The Indian Evidence Act, 1872.
Section 65 B of The Indian Evidence Act, 1872.

Prominent Arguments by the Advocates:

1. **Appellant**
   (a) It was argued that the video cassette's authenticity was affirmed by the CFSL, the cassette was deemed original, thus should be considered admissible as primary evidence, contrary to the trial court's decision.
   (b) It was contented that the objections regarding the certificate under Section 65B of the Indian Evidence Act cannot be raised at this stage, citing the principle that such objections must be raised at the material time.
2. **Respondent**
   (a) It was argued that the video recording in question lacked admissibility as both primary and secondary evidence, as per Section 62 of the Indian Evidence Act, due to the prosecution's failure to prove its originality and suspicions of tampering. Additionally, the absence of a Section 65B certificate further hindered its admissibility as secondary evidence.

47 Corruption Case No. 20/2013 arising out of FIR No. 383/2007 registered at PS Dabri.

(b) It was contended that tape recorded conversations could only serve as corroborative evidence of actual conversations testified by any party involved, yet without material evidence of such conversations, the tape recordings lacked reliability.

Judgement:

1. It was stated that the analyses of the learned special judge is a *"possible view"*. The analysis being that *"there is a possibility that the recording was stopped in between or some portions were deleted"*.
2. It was pointed that even though it is asserted that a new cassette is used, the same was contrary to the findings of ETC channel clippings in the Evidence.
3. It was held that the video cassette is not admissible. The same was supported by the fact that:
   (a) It was not supported by certificate under Section 65B of Indian Evidence Act, 1872.
   (b) It was not proved that the recording is original.

Final Decision:

1. The appeal was dismissed and disposed.
2. Bail Bonds furnished by respondents stand discharged.

LET'S RECALL

Sr. No.	*Name*	*Issue of Interest*	*Electronic Record in Question*	*Held*	*Current Status*	*Remarks*
1.	**R.M. MALKANI V. STATE OF MAHARASHTRA** 1972 | SC | 2 Judge Bench | (1973) 1 SCC 471	Whether tape-recorded telephone conversation admissible as evidence?	Tape-Recording	A tape-recorded conversation is admissible in the court of law, provided • It is relevant to the matter, • The voice is identified, • The authenticity of tape-recorded conversation is proved. It is also comparable to a photograph of a relevant incident.	—	Discusses admissibility of an Electronic Record as evidence even before the term was defined in the Indian Laws.
2.	**ZIYAUDDIN BURHANUDDIN BUKHARI V. BRIJMOHAN RAMDASS MEHRA & ORS.** 1975 | SC | 3 Judge Bench | (1976) 2 SCC 17	Whether tape-recordings of election speeches admissible as evidence?	Tape-Recording	Tape records of speeches are "documents" under Section 3 of the Evidence Act and stand on no different footing than photographs. There could be no more direct or better evidence of it than its taperecord, if its authenticity is duly established.	—	Discusses admissibility of an Electronic Record as evidence even before the term was defined in the Indian Laws.
3.	**STATE (NCT OF DELHI) V. NAVJOT SANDHU** 2005 | SC | 2 Judge Bench | 8 (2005) 11 SCC 600	Whether the electronic records (phone call records) are admissible as evidence?	Phone Call Records	Regardless of whether or not the requirements of Section 65-B are met, the requirement of certificate under Section 65B is not always mandatory. Electronic evidence be admitted u/s 63 & 65 as secondary evidence without certificate.	Overruled by *Anvar P.V.* v. *P.K. Basheer* & Ors.	Following this decision, high courts all around the nation began to accept other types of proof in place of certificates, such as oral testimony, which inadvertently lowered the bar for electronic evidence.
4.	**TUKARAM S. DIGHOLE V. MANIKRAO SHIVAJI KOKATE** 2010 | SC | 2 Judge Bench | (2010) 4 SCC 329	Whether mere production of an audio cassette (assuming it to be a certified copy) is conclusive of the fact that its content are true and correct?	VHS Cassette	Authenticity of VHS Cassette was not established by the petitioner. Held that technology based on audio and videotapes might serve as a significant piece of evidence but are more prone to tampering and alteration, therefore needs to be treated with caution.	—	—

Sr. No.	*Name*	*Issue of Interest*	*Electronic Record in Question*	*Held*	*Current Status*	*Remarks*
5.	**ANVAR P.V. V. P.K. BASHEER & ORS.** 2014 | SC | 3 Judge Bench | (2014) 10 SCC 473	Interpretation of Section 65B—Admissibility of Electronic Evidence	CD	Section 65B is a code in itself with respect to Electronic Records. The requirement to produce certificate is necessary. Sections 63 and 65 do not apply to electronic records as secondary evidence.	Upheld by *Arjun Panditrao Khotkar* v. *Kailash Kushanrao Gorantyal*	—
6.	**TOMASO BRUNO AND ANR. V. STATE OF UTTAR PRADESH** 2015 | SC | 3 Judge Bench | (2015) 7 SCC 178	Rule of Best Evidence in the age of Science & Technology	CCTV Footage	Electronic documents are accepted as material evidence in the strictest sense. CCTV footage would have been best and a strong piece of evidence.	Overruled by *Arjun Panditrao Khotkar* v. *Kailash Kushanrao Gorantyal*	As per the Arjun Panditrao Judgement, the court's view of admissibility of CCTV footage as secondary evidence u/s 65 of the act was incorrect statement of law followed by citing Navjot Sandhu judgment that was overruled by the Anvar PV judgement (also not referred).
7.	**SONU ALIAS AMAR V. STATE OF HARYANA** 2017 | SC | 2 Judge Bench | (2017) 8 SCC 570	Objection regarding mode/method of proof of electronic record at appellate stage	CDRs (Call Detail Records)	Objections cannot be permitted to be raised at appellate stage regarding mode/method of proof of electronic record.	—	Such objection was raised for the first time before the hon'ble Supreme Court.
8.	**SHAFHI MOHAMMAD V. STATE OF HIMACHAL PRADESH** 2018 | SC | 2 Judge Bench | (2018) 2 SCC 801	Whether use of videography be allowed at scene of crime? Applicability of Section 65B	Video	Sections 65A and 65B are just clarifying and procedural in character, and cannot be considered a comprehensive regulation on the issue. Requirement of Certificate not neccessary, evidence could be produced under Sections 63 and 65.	Overruled by *Arjun Panditrao Khotkar* v. *Kailash Kushanrao Gorantyal*	The judgement disregarded a well-established precedent and erred by overruling a higher judge bench decision. It violated some procedural norms established by common law.

Sr. No.	Name	Issue of Interest	Electronic Record in Question	Held	Current Status	Remarks
9.	**ARJUN PANDITRAO KHOTKAR V. KAILASH KUSHANRAO GORANTYAL** 2020 | SC | 3 Judge Bench | (2020) 7 SCC 1	Interpretation of Section 65B—Admissibility of Electronic Evidence	CCTV Footage	Sections 65A & 65B form a comprehensive code governing the admissibility of electronic records as evidence. The certificate shall be necessarily produced when the e-evidence is secondary, and it can be produced anytime during the proceeding.	—	—
10.	**RAVINDER SINGH ALIAS KAKU V. STATE OF PUNJAB** 2022 | SC | 2 Judge Bench | (2022) 7 SCC 581	Whether the call records would be admissible under S 65A and 65B with-out compliance of requirement of certificate?	Call Records	Section 65 B(4) is a mandatory requirement of law and that oral evidence cannot suffice in place of such certificate.	—	—
11.	**STATE OF KARNATAKA VS. T. NASEER AND ORS** 2023 | SC | 2 Judge Bench | MANU/SC/1217/2023	Interpretation of Section 65B—Requirement of Certicate	Electronic Devices	Certificates under Section 65B are essential when electronic evidence is presented as secondary evidence but are not obligatory when the electronic record is employed as primary evidence.	—	—
12.	**STATE (GNCT OF DELHI) V. NETRAPAL SINGH & ORS** 2024 | High Court of Delhi | MANU/DE/0135/2024	Whether the video recording be admissible, considering concerns over its originality, potential tampering, and lack of the Certificate as under Section 65B?	Video Cassette	The video cassette was neither supported by certificate u/s 65B nor it was proved to be original and authentic. Thus, the video cassette was held to be inadmissible.	—	—

Knowledge Check

A. CHOOSE THE CORRECT OPTION

1. What was the main contention regarding the admissibility of VCD (E-Evidence) in the judgement of *Arjun Panditrao Khotkar* v. *Kailash Kushanrao Gorantyal*?
   (a) The absence of a certificate under Section 65B(4)
   (b) The reliability of the VCD content
   (c) The procedural compliance with Section 65A
   (d) All of the above
2. Under Section 65B(2) of the Indian Evidence Act, what conditions must a computer meet to be considered a reliable source?
   (a) It must be brand new
   (b) It must be the one regularly used to save information
   (c) It must be connected to the internet
   (d) It must have a high processing speed
3. In the *R.M. Malkani* v. *State of Maharashtra* case, what was the main issue regarding admissibility?
   (a) Admissibility of VHS Cassettes
   (b) Admissibility of Electronic Records
   (c) Admissibility of Tape-Recorded Evidence
   (d) Admissibility of Written Statements
4. What led to the Trial Court's refusal to admit the CFSL report as evidence in the *State of Karnataka* v. *T. Naseer and Ors.*?
   (a) Lack of relevance of electronic evidence
   (b) Absence of a Section 65B certificate
   (c) Inadmissible forensic findings
   (d) Delay in Submission
5. Who typically issues the Section 65B certificate for electronic evidence?
   (a) Investigating officer
   (b) Forensic expert
   (c) Judge overseeing the case
   (d) Relevant government authority

Answer Key

1. (a) The absence of a certificate under Section 65B(4).
2. (b) It must be the one regularly used to save information.

3. (c) Admissibility of Tape-Recorded Evidence
4. (b) Absence of a Section 65B certificate.
5. (d) Relevant government authority

B. ANSWER THE FOLLOWING

1. What specific legal framework governs the admissibility of electronic evidence in India?
2. According to the judgment in *Anvar P.V.* v. *P.K. Basheer & Ors.*, what conditions must be satisfied for the admissibility of electronic records? Explain in detail.
3. What role did electronic evidence, particularly call records, play in the trial in the case of *Ravinder Singh alias Kaku* v. *State of Punjab*?
4. In the *Sonu alias Amar* v. *State of Haryana case*, what objection did the appellant raise regarding electronic records, and how did the court respond? Give a detailed analysis.
5. In the case of *R.M. Malkani* v. *State of Maharashtra*, what was the main issue regarding admissibility? Explain.
6. How did the court justify the admissibility of CD or VCDs in the absence of a certificate under Section 65B(4) in the case of Arjun *Panditrao Khotkar* v. *Kailash Kushanrao Gorantyal*?
7. How did the Supreme Court address the applicability of Section 65B(4) of the Evidence Act in *Shafhi Mohammad* v. *State of Himachal Pradesh*?
8. What was the significance of the judgment in *Tomaso Bruno* v. *State of UP* regarding electronic evidence?
9. How did the Supreme Court address the delay in producing the Section 65B certificate in State of *Karnataka* v. *T. Naseer and Ors.*?
10. In the case of *Arjun Panditrao Khotkar* v. *Kailash Kushanrao Gorantyal*, what was the main contention regarding the admissibility of VCD (E-Evidence)?
11. In what circumstances are tape records considered admissible as evidence according to the court's decision in Ziyauddin *Burhanuddin Bukhari* v. *Brijmohan Ramdass Mehra & Ors.*?
12. How did the court address the admissibility of the VHS cassette as evidence, and what legal principles were considered in *Tukaram S. Dighole* v. *Manikrao Shivaji Kokate*?

CHAPTER 7

International Perspectives on Electronic Evidence Laws

A Glimpse into the Chapter

Electronic Evidence Landscape in:
- USA
- UK
- New Zealand
- Australia

As the world becomes more interconnected and reliant on digital technology, understanding international perspectives on electronic evidence laws becomes crucial. This chapter sheds light on the key legal principles, regulations, and landmark cases that shape the admissibility and treatment of electronic evidence in these diverse jurisdictions. This explores the complex landscape of electronic evidence laws in four distinct regions: the United States, the United Kingdom, New Zealand, and Australia. Each of these regions has developed a unique legal framework to address the challenges and opportunities presented by electronic evidence in the courtroom.

7.1 EVIDENCE LAWS IN USA

Electronic evidence laws in the United States pertain to the admissibility, collection, preservation, and presentation of digital evidence in legal proceedings. These laws are primarily governed by federal and state rules of evidence and regulations.

1. **Federal Rules of Evidence**

   They are a set of guidelines that federal courts in the U.S. follow when it comes to the admissibility of evidence in legal proceedings. Following are the key rules regarding electronic evidence[1]:

   (a) **Rule 901—Authentication[2]**

   This rule requires that evidence, including electronic evidence, must be authenticated before it can be admitted in court. Authentication[3] involves showing that the evidence is what it purports to be.

   (b) **Rule 902—Self-Authentication[4]**

   Rule 902 lists examples of evidence that are considered self-authenticating[5]. The rule includes provisions for various forms of evidence such as public documents, foreign records, certified copies, official publications, newspapers, and more. Certain electronic evidence, including certified copies of records generated by electronic processes or systems, or self-authenticated data generated by electronic processes, are explicitly addressed under the Rule.

   (c) **Rule 803—Exceptions to the Rule Against Hearsay[6]**

   The rule outlines exceptions to the hearsay rule, allowing the admission of certain statements, records, or data compilations[7], regardless of whether the declarant is available as a witness. These exceptions aim to admit reliable evidence, such as statements in ancient documents,

---

1 Federal Rules of Evidence, Retrieved from: https://www.law.cornell.edu/rules/fre

2 Rule 901, Federal Rules of Evidence, Retrieved from: HTTPs://www.law.cornell.edu/rules/fre/rule_901.

3 Authentication, in this context, involves presenting sufficient evidence to demonstrate that the proffered evidence is indeed what it is claimed to be. In the case of electronic evidence, this may include establishing the origin, integrity, and accuracy of digital information, ensuring that it has not been tampered with or altered.

4 Rule 902, Federal Rules of Evidence, Retrieved from: https://www.law.cornell.edu/rules/fre/rule_902.

5 Evidence that can be admitted without requiring additional proof of authenticity.

6 Rule 803, Federal Rules of Evidence, Retrieved from: https://www.law.cornell.edu/rules/fre/rule_803.

7 As defined under Notes of Advisory Committee on the Proposed Rules the term "data compilation" is expansively defined to include any means of storing information, extending beyond traditional written or documentary forms to encompass electronic computer storage and other formats. This definition reflects the evolving nature of information storage and retrieval, recognizing the admissibility of electronic data in legal proceedings.

judgments involving personal or family history, reputation evidence, and judgments of previous convictions, while safeguarding against potential issues of trustworthiness.

(d) **Rule 1001—Definitions That Apply to This Article**[8]

This rule outlines key definitions relevant to evidence in modern legal contexts. They acknowledge the evolving nature of information storage, encompassing electronic data, and establish criteria for the admissibility of electronic evidence. The definition of "writing", "recording," and "photograph" explicitly acknowledges the inclusion of electronic equivalents[9]. The definition of an "original" for electronically stored information includes any sight-readable output that accurately reflects the information. The term "duplicate" is defined to cover counterparts produced by various methods, including electronic processes.

2. **Case Laws**

Many court decisions have shaped the admissibility and treatment of electronic evidence. These cases often address issues like the authentication of digital evidence, chain of custody, and privacy concerns. Below listed are a few landmark judgments:

(a) **Zubulake v. UBS Warburg (2003)**[10]

Zubulake is a significant e-discovery case that emphasized the duty to preserve electronic evidence in civil litigation.

---

8 Rule 1001, Federal Rules of Evidence, Retrieved from: https://www.law.cornell.edu/rules/fre/rule_1001. The rule states: Definitions that Apply to this Article
In this article:

(a) A "writing" consists of letters, words, numbers, or their equivalent set down in any form.
(b) A "recording" consists of letters, words, numbers, or their equivalent recorded in any manner.
(c) A "photograph" means a photographic image or its equivalent stored in any form.
(d) An "original" of a writing or recording means the writing or recording itself or any counterpart intended to have the same effect by the person who executed or issued it. For electronically stored information, "original" means any printout—or other output readable by sight—if it accurately reflects the information. An "original" of a photograph includes the negative or a print from it.
(e) A "duplicate" means a counterpart produced by a mechanical, photographic, chemical, electronic, or other equivalent process or technique that accurately reproduces the original.

9 The Notes of Advisory Committee on Proposed Rules emphasize the rule's adaptation to modern data storage methods, such as computers and photographic systems, while amendments explicitly include "video tapes" in the definition of "photographs."

10 *Zubulake* v. *UBS Warburg LLC,* 220 F.R.D. 212 (S.D.N.Y. 2003).

It underscores the importance of electronic documents in modern legal proceedings. The case established that parties have an obligation to take reasonable steps to preserve and produce electronic documents. This is critical for ensuring the integrity and admissibility of electronic evidence in civil cases.[11]

(b) **United States v. Mitra (2005)**[12]

This case highlighted the importance of maintaining a clear chain of custody for electronic evidence. The chain of custody is essential for establishing the authenticity and reliability of the evidence. Courts require a documented record of who had control of the evidence from the time it was collected to its presentation in court. A well-maintained chain of custody is a key factor in the admissibility of electronic evidence.

(c) **Lorraine v. Markel American Insurance Co. (2007)**[13]

Lorraine is a significant case for the admissibility of electronic evidence related to digital communications. It emphasizes the importance of demonstrating the authenticity and relevance of electronic evidence, particularly in the context of emails. The case discusses the use of metadata and other factors in determining admissibility, showing that the proper handling and explanation of electronic evidence are essential.[14]

(d) **United States v. Vosburgh (2010)**[15]

In this case, the court's ruling highlighted that instant messages and chat logs could be admissible as long as their authenticity and relevance were established. This decision underscores the importance of authentication and

11 *Zubulake* v. *UBS Warburg* I: E-Discovery Case Law, Exterro Blog, Retrieved from: https://www.exterro.com/blog/zubulake-v-ubs-warburg-i.

12 *United States* v. *Mitra-Hernandez*, 405 F.3d 492 (7th Cir. 2005).

13 *Lorraine* v. *Markel American Ins. Co.*, 241 F.R.D. 534 (D. Md. 2007) .

14 *Grimm, Hon. Paul W.; Ziccardi, Michael* v. *Esq.*; and *Major, Alexander W. Esq.* (2009) "Back to the Future:
Lorraine v. *Markel American Insurance Co.* and New Findings on the Admissibility of Electronically Stored Information," *Akron Law Review*: Vol. 42: Iss. 2 , Article 2. Available at: http://ideaexchange.uakron.edu/akronlawreview/vol42/iss2/2.

15 *U.S.* v. *Vosburgh*, 602 F.3d 512 (3d Cir. 2010).

relevance when presenting electronic evidence, particularly in cases involving digital communications.[16]

(e) **United States v. Jones (2012)**[17]

The Jones case addressed issues of electronic surveillance, particularly the use of GPS tracking devices. The Supreme Court's ruling that attaching a GPS device to a suspect's vehicle constituted a search under the Fourth Amendment has implications for the admissibility of electronic evidence obtained through certain surveillance methods. It underscores the need for legal authorization and adherence to constitutional protections when collecting electronic evidence.[18]

(f) **Riley v. California (2014)**[19]

Riley is a landmark case concerning the admissibility of electronic evidence obtained from mobile devices. The Supreme Court held that law enforcement generally needs a warrant to search the digital contents of a cell phone seized from an individual's person or vehicle. This decision highlights the need for proper legal procedures when dealing with electronic evidence, especially from mobile devices.[20]

7.2 EVIDENCE LAWS IN UK

In the United Kingdom, the admissibility of electronic evidence is governed by various laws and regulations.

1. **Civil Evidence Act 1995**[21]

   The Act addresses the admissibility of electronic evidence in civil proceedings in the UK. Provisions of this act include:

---

16 *United States* v. *Vosburgh*, 602 f.3d 512 (2010): Case brief. Rederived from: https://www.quimbee.com/cases/united-states-v-vosburgh.

17 *United States* v. *Jones*, 565 U.S. 400 (2012).

18 *United States v. Jones*, Rederived from: https://epic.org/documents/united-states-v-jones/.

19 *Riley* v. *California*, 573 U.S. 373 (2014).

20 Global Freedom of Expression. *Riley* v. *California*, Retrieved from: https://globalfreedomofexpression.columbia.edu/cases/riley-v-california-2/.

21 Civil Evidence Act 1995, Retrieved from: https://www.legislation.gov.uk/ukpga/1995/38/contents.

(a) **Section 5—Competence and Credibility**[22]

This section provides for the admissibility of statements contained in electronic documents. To be admissible, these statements should be produced by a reliable process and in its primary form. The reliability of the electronic document's creation and storage process is a crucial consideration.

2. **Criminal Justice Act 2003**[23]

The Criminal Justice Act 2003 introduced provisions related to the admissibility of hearsay evidence in criminal trials in England and Wales[24]. Hearsay evidence refers to statements made outside of court that are offered as proof of the truth of the matters asserted. Electronic records and communications can fall into the category of hearsay evidence if they contain statements made outside of court. Here are a few key provisions:

(a) **Section 114—Admissibility of Hearsay Evidence**[25]

The section introduces exceptions to the hearsay rule for specific types of electronic evidence, like computer records used in business activities. Additionally, the provision introduces specific exceptions to the hearsay rule for certain types of electronic evidence, allowing them to be admitted as evidence without needing direct witness testimony.[26]

(b) **Section 115—Statements and Matters Stated**[27]

The provision allows courts to admit computer-generated evidence, such as financial records or traffic camera footage, if certain conditions are met, like:

(i) *Adequacy of a Computer System:* The computer system used to generate the evidence must have been properly maintained and operated.

---

22 Section 5, Civil Evidence Act 1995, Retrieved from: https://www.legislation.gov.uk/ukpga/1995/38/section/5.

23 Criminal Justice Act 2003, Retrieved from: https://www.legislation.gov.uk/ukpga/2003/44/contents.

24 The Criminal Justice Act 2003 only applies to England and Wales. Scotland and Northern Ireland have their own laws governing the admissibility of hearsay evidence.

25 Section 114, Criminal Justice Act 2003, Retrieved from: https://www.legislation.gov.uk/ukpga/2003/44/section/114.

26 Notes on Hearsay Evidence by Ezekiel Chin, Retrieved from: https://www.scribd.com/document/575549148/Hearsay-Evidence.

27 Section 115, Criminal Justice Act 2003, Retrieved from: https://www.legislation.gov.uk/ukpga/2003/44/section/115.

(ii) *Proper Records Kept:* Records must be kept that show how the evidence was generated and preserved.

(iii) *Authentication:* The court must be satisfied that the evidence is authentic and has not been tampered with.[28]

(c) **Section 116—Cases Where a Witness is Unavailable**[29]

The provision deals with the admissibility of electronic communications, such as emails, text messages, and social media posts. Similar to Section 115, the court must be satisfied that the communication is authentic and that it originated from the source claimed. Other factors the court may consider include the timing of the communication, the context in which it was made, and the relationship between the sender and recipient.[30]

3. **Data Protection Act 2018**[31]

The Data Protection Act 2018 is essential for cases involving data breaches, privacy violations, and the handling of electronic records. Relevant provisions include:

(a) **Schedule 22**

This schedule covers the processing of personal data for law enforcement purposes. It includes rules regarding the processing of electronic data and the admissibility of evidence in cases related to data protection breaches. It mainly focuses on the conditions under which law enforcement agencies can process personal data, including electronic data.

4. **Civil Procedure Rules (CPR)**[32]

The Civil Procedure Rules in the UK provide rules and guidelines for civil litigation, including the handling of electronic evidence. Key provisions include:

---

28 Notes on Hearsay Evidence by Ezekiel Chin, Retrieved from: https://www.scribd.com/document/575549148/Hearsay-Evidence.

29 Section 116, Criminal Justice Act 2003, Retrieved from: https://www.legislation.gov.uk/ukpga/2003/44/section/116.

30 Notes on Law of Evidence by Jay Khan, Retrieved from: https://www.studocu.com/en-gb/document/university-of-bradford/law-of-evidence/law-of-evidence-assignment/15724062.

31 Data Protection Act 2018, Retrieved from: https://www.legislation.gov.uk/ukpga/2018/12/contents/enacted.

32 The Civil Procedure Rules 1998, Retrieved from: https://www.legislation.gov.uk/uksi/1998/3132/contents.

(a) **Part 31**

It focuses on the disclosure and inspection of electronic documents during the discovery phase of civil cases. It outlines the procedures for handling electronic evidence in civil litigation, emphasizing the need for proportionality and cooperation among parties. This means that parties should only disclose what is relevant and proportionate to the case, and they should cooperate in good faith to exchange electronic documents efficiently and effectively. Part 31 applies to all electronic documents that are relevant to the issues in dispute, regardless of their format or storage medium.

The rules outline specific procedures for disclosing electronic documents, including:

(i) *Providing a list of Electronic Documents:* Each party must provide a list of all relevant electronic documents in their possession or control.

(ii) *Specifying the Format:* Parties must specify the format in which the documents will be disclosed, taking into account the recipient's ability to access and view them.

(iii) *Cost Considerations:* The rules encourage parties to consider the cost of disclosure when deciding what documents to disclose.

(iv) *Confidentiality:* The rules also address issues of confidentiality and data protection when dealing with sensitive electronic documents.[33]

5. **Criminal Procedure Rules (CrimPR)**[34]

The Criminal Procedure Rules in the UK address the presentation and handling of evidence in criminal proceedings. Relevant provisions include:

(a) **Rule 9**

It covers evidence, including electronic evidence, in criminal trials. It sets out the procedures for presenting evidence, including the admissibility of electronic records, in criminal cases.

---

33 Coulson, P. J., Fontaine, S. M., & Sorabji, J. (Eds.). (2023). The White Book Service 2023: Civil Procedure Volumes 1 & 2. Sweet & Maxwell. (ISBN 9780414112339).

34 The Criminal Procedure Rules 2020, Retrieved from: https://www.legislation.gov.uk/uksi/2020/759/contents/made.

6. **Regulation of Investigatory Powers Act 2000 (RIPA)**[35]
   RIPA regulates the interception of communications and surveillance in the UK, which may involve electronic evidence. Part II of RIPA covers the interception of communications and the use of electronic surveillance methods. The act provides legal authorization for surveillance and sets out rules for the admissibility of evidence obtained through authorized surveillance. RIPA sets out strict criteria for when public authorities can conduct surveillance, requiring prior authorization from appropriate bodies like the Investigatory Powers Tribunal. Admissibility depends on legality, i.e. evidence obtained through unlawful surveillance may not be admissible in court.
7. **Common Law Principles**
   Common law principles in the UK provide a foundation for the admissibility of all types of evidence, including electronic evidence. These principles emphasize fairness, reliability, and relevance. Courts assess whether electronic evidence meets these criteria when determining its admissibility.
8. **Case Laws**
   UK courts have handed down decisions that set precedents and principles regarding the admissibility of electronic evidence. These cases often involve complex issues such as authentication, reliability, and the legal procedures for presenting electronic evidence.
   (a) **R v. Bowden (2005)**[36]
   R v. Bowden dealt with child pornography charges and focused on the admissibility of downloaded and printed indecent images. In this case, the court addressed the admissibility of electronic evidence in criminal proceedings. The case highlighted the need for the proper handling and authentication of electronic records and communications. The court emphasized that the reliability and integrity of electronic evidence were key factors in its admissibility.
   (b) **Barclays Bank Plc v. Devonport Royal Dockyard Ltd (2008)**[37]
   The case dealt with a complex commercial dispute relating to a loan agreement, where electronic evidence played a

35 Regulation of Investigatory Powers Act 2000, Retrieved from: https://www.legislation.gov.uk/ukpga/2000/23/contents.

36 *R* v. *Bowden* [2005] EWCA Crim 2270.

37 *Barclays Bank Plc* v. *Devonport Royal Dockyard Ltd* [2008] EWHC 1509 (QB).

significant role. It highlighted the challenges and procedures associated with presenting electronic evidence in high-value litigation. The case focused on issues like email trails, deleted emails, and the burden of proof in relation to electronic disclosure. The court dismissed Devonport Royal Dockyard's argument concerning certain email evidence, ruling that it was admissible despite claims of deletion and subsequent recovery.

The court emphasized the importance of proportionality and reasonableness in electronic disclosure, balancing the cost and burden of disclosure with the relevance and importance of the evidence. The court clarified the burden of proof for electronic disclosure, placing the onus on the party seeking to withhold evidence to demonstrate valid reasons.

(c) **Serco Ltd v. Secretary of State for Defence (2011)**[38]
This case addressed the disclosure of electronic documents in the context of public law and procurement. It underscored the importance of the electronic discovery process and the admissibility of electronically stored information in administrative and public law cases. The court ordered the Secretary of State to disclose additional electronic documents related to the Ministry of Defence tender process. The court confirmed the importance of proportionality in electronic disclosure, balancing the cost and burden of disclosure with the relevance of the documents to the case. The court clarified the principles for determining the admissibility of ESI[39], emphasizing the need for proper authentication and ensuring the evidence is unaltered and reliable.

(d) **Dawson-Damer & Ors v. Taylor Wessing LLP (2017)**[40]
This civil case emphasized the relevance of electronic evidence and data protection issues. It highlighted the importance of complying with data protection laws when handling electronic evidence, especially in the context of data subject access requests. The Court of Appeal ruled in favor of the data subjects, finding that Taylor Wessing LLP could not rely on legal professional privilege to withhold all personal data requested by the beneficiaries, and needed to

38 *Serco Ltd v. Secretary of State for Defence* [2011] EWHC 348 (TCC).
39 Electronically Stored Information.
40 *Dawson-Damer & Ors v. Taylor Wessing LLP* [2017] EWCA Civ 74.

balance legal professional privilege with the data subjects' right to access their personal data under the DPA 1998.

(e) **Alibi v. CitySprint UK Ltd (2019)**[41]

This case revolved around electronic evidence related to gig economy workers' employment status. It demonstrated the significance of electronic evidence in establishing the terms of a worker's employment and the need to ensure the admissibility of such evidence. The court focused on how the electronic evidence reflected the nature of the working relationship between the drivers and CitySprint, considering factors like control, supervision, and integration into the business.

7.3 EVIDENCE LAWS IN NEW ZEALAND

In New Zealand, the admissibility of electronic evidence is governed by a combination of laws and legal principles. Here are some key legal considerations and statutes that address the admissibility of electronic evidence in New Zealand:

1. **Evidence Act 2006**[42]

   The Evidence Act 2006 is the primary piece of legislation that governs the admissibility of evidence in New Zealand. It contains provisions related to electronic evidence, including definitions of documents, admissibility criteria, and the principles of relevance and reliability. Key Provisions of the act include:

   (a) **Section 4(1)—Interpretation—Document**[43]

   Defines a "document" to include any data held in electronic form.[44]

---

41 *Alibi* v. *CitySprint UK Ltd* [2019] ETCA 68.

42 Evidence Act 2006 No 69, Retrieved From: https://www.legislation.govt.nz/act/public/2006/0069/latest/DLM393463.html.

43 Section 4(1), Evidence Act 2006 No 69, Retrieved From: https://www.legislation.govt.nz/act/public/2006/0069/latest/DLM393471.html.

44 It states: document means—

(a) any material, whether or not it is signed or otherwise authenticated, that bears symbols (including words and figures), images, or sounds or from which symbols, images, or sounds can be derived, and includes—

(i) a label, marking, or other writing that identifies or describes a thing of which it forms part, or to which it is attached;

(ii) a book, map, plan, graph, or drawing;

(iii) a photograph, film, or negative; and

(b) information electronically recorded or stored, and information derived from that information.

(b) **Section 106—Video Record Evidence**[45]

The provision outlines the procedures for the admissibility of video record evidence in criminal proceedings. It stipulates that video record[46] evidence may be presented as an alternative method for witnesses to give evidence in chief, emphasizing compliance with regulations. The section details the conditions for providing copies of video records, exceptions in specific cases, and the Judge's authority to order provisions in exceptional circumstances. The opportunity for all parties to make submissions on admissibility, the process for the defendant's objection, and the Judge's discretion to exclude material are also specified. Additionally, the section addresses the admissibility of video records despite non-strict observance of regulations and the exclusion of subsections for Crown lawyers.

(c) **Section 136—Proof of Signatures on Attested Documents**[47]

It expressly allows for the verification of signatures, execution, or attestation through any satisfactory means including digital methods. Notably, the section eliminates the obligatory requirement of calling an attesting witness, irrespective of the signing method, thereby acknowledging the validity of digital signatures within the realm of electronic evidence.

2. **Electronic Transactions Act 2002**[48]

The Electronic Transactions Act 2002 provides a legal framework for electronic transactions in New Zealand. The Act aids businesses, individuals, and the government in electronically administering commercial transactions, granting specific electronic information an equivalent legal standing to that of paper-based information.[49]

---

45 Section 106, Evidence Act 2006 No. 69, Retrieved from: https://www.legislation.govt.nz/act/public/2006/0069/latest/DLM393939.html?search=sw_096be8ed81de303d_electronic_25_se&p=1&sr=1.

46 Section 106 (10): A video record includes a reference to the person being given access to the video record, for example, being given access to an electronic copy of the video record through an Internet site.

47 Section 136, Evidence Act 2006 No. 69, Retrieved from: https://www.legislation.govt.nz/act/public/2006/0069/latest/DLM393978.html?search=sw_096be8ed81de303d_digital_25_se&p=1&sr=0.

48 Electronic Transactions Act 2002 (2002 No. 35), Retrieved from: https://www.legislation.govt.nz/act/public/2002/0035/latest/DLM154185.html.

49 What is the Electronic Transactions Act in New Zealand? Retrieved from: https://legalvision.co.nz/commercial-contracts/electronic-transactions-act/.

The said act was repealed by the Contract and Commercial Law Act 2017[50].

3. **The Contract and Commercial Law Act 2017**[51]
   Part 4 of the Act deals with Electronic Transactions. This part aims to make it easier to use electronic technology by making things clearer. It reduces uncertainty about the legal effect of electronic information and when electronic messages are sent and received. It also allows electronic technology to be *functionally equivalent* to certain paper-based legal requirements if it does the same job[52]. Its key provisions include:
   (a) **Section 209—Interpretation**[53]
   The section provides an inclusive definition of the term Electronic[54]. It defines a Data storage device[55] as any article or device capable of reproducing information. Further, amongst others, it defines electronic communication[56], electronic signature[57], information[58], and information system[59].
   (d) **Section 211—Validity of Information**[60]
   This provision clarifies that information is not deprived of legal validity solely because it exists in electronic form or is

50 Contract and Commercial Law Act 2017 (2017 No. 5).

51 Contract and Commercial Law Act 2017 (2017 No 5) Retrieved from: https://www.legislation.govt.nz/act/public/2017/0005/latest/DLM6844033.html?search=sw_096be8ed81df44a8_electronic_25_se&p=1#DLM6844433.

52 Section 207, Contract and Commercial Law Act 2017 (2017 No 5), Retrieved from: https://www.legislation.govt.nz/act/public/2017/0005/latest/DLM6844429.html?search=sw_096be8ed81df44a8_transaction_25_se&p=1&sr=7.

53 Section 209, Contract and Commercial Law Act 2017 (2017 No 5), Retrieved from: https://www.legislation.govt.nz/act/public/2017/0005/latest/DLM6844433.html?search=sw_096be8ed81df44a8_electronic_25_se&p=1&sr=6.

54 Electronic includes electrical, digital, magnetic, optical, electromagnetic, biometric, and photonic.

55 Data storage device means any article or device (for example, a disk) from which information is capable of being reproduced, with or without the aid of any other article or device.

56 Electronic communication means a communication by electronic means.

57 Electronic signature, in relation to information in electronic form, means a method used to identify a person and to indicate that person's approval of that information

58 Information includes information (whether in its original form or otherwise) that is in the form of a document, a signature, a seal, data, text, images, sound, or speech.

59 As defined u/s 213(2) information system means a system for producing, sending, receiving, storing, displaying, or otherwise processing electronic communications.

60 Section 211, Contract and Commercial Law Act 2017 (2017 No. 5), Retrieved from: https://www.legislation.govt.nz/act/public/2017/0005/latest/DLM6844453.html?search=sw_096be8ed81df44a8_electronic_25_se&p=1.

conveyed through electronic communication. It emphasizes that the legal effect of information is not diminished merely because it is presented or referred to as electronic.

(c) **Section 235—Legal Requirement to Provide Access to Information that is in Electronic Form**[61]

To comply with a legal obligation to provide access to electronic information, it can be done by presenting the information in paper or any non-electronic form. If ensuring the integrity of the information in non-electronic form is uncertain, the responsible party must inform everyone who needs access and, if asked, provide the information in electronic form. Electronic access is acceptable if it reliably maintains the information's integrity, considering the purpose and circumstances, and if the concerned party agrees to access it electronically.

4. **Criminal Disclosure Act 2008**[62]

The Criminal Disclosure Act 2008 is relevant in criminal cases and governs the disclosure of evidence, including electronic evidence, to the defense. It ensures that electronic evidence is properly disclosed to all parties involved in a criminal case. The act does not limit the information based on the form it is contained in or the medium it is recorded in, it explicitly includes information recorded in electronic form.[63]

5. **Search and Surveillance Act 2012**[64]

The Search and Surveillance Act 2012 regulates the search and surveillance powers of law enforcement agencies. It is relevant to

---

61 Section 235, Contract and Commercial Law Act 2017 (2017 No 5), Retrieved from: https://www.legislation.govt.nz/act/public/2017/0005/latest/DLM6844485.html?search=sw_096be8ed81df44a8_electronic_25_se&p=1.

62 Criminal Disclosure Act 2008 (2008 No 38), Retrieved from: https://www.legislation.govt.nz/act/public/2008/0038/latest/DLM1378803.html?search=sw_096be8ed81de6247_electronic_25_se&p=1.

63 Section 6(2), Criminal Disclosure Act 2008 (2008 No 38), Retrieved from: https://www.legislation.govt.nz/act/public/2008/0038/latest/DLM1378813.html?search=sw_096be8ed81de6247_electronic_25_se&p=1&sr=0.
The provision States:
6(2) In this Act, a reference to information means any recorded information—
(a) in whatever form it is contained, for example, in a report, statement, list, or interview; and
(b) in whatever medium it is recorded, for example, in hard copy, electronic form, or as a sound or visual recording.

64 Search and Surveillance Act 2012 (2012 No. 24), Retrieved from: https://www.legislation.govt.nz/act/public/2012/0024/latest/DLM2136536.html.

cases where electronic evidence is obtained through search and seizure. The act sets out procedures and safeguards for collecting electronic evidence during investigations, hence formalizing not just electronic evidence but digital forensic procedures as law. The interpretation clause[65] defines various key terms including:

(a) "Access" to a computer system includes various actions like giving instructions, communicating with, storing or receiving data, or using any resources of the computer system.

(b) "Access information" involves things like codes, passwords, and encryption keys that enable access to a computer system or other data storage devices.

(c) "Computer system" refers to a single computer, interconnected computers, communication links, or a combination of interconnected computers with communication links.

(d) "Interception device" is any electronic or mechanical instrument used to intercept or record private communications, excluding hearing aids.

(e) "Remote access search" is searching something like an Internet data storage facility without a physical address.

(f) "Surveillance device" includes interception, tracking, or visual surveillance devices, which include an electronic device.

(g) "Tracking device" helps determine the location of a thing or person.

(h) "Visual surveillance device" observes or records private activities, excluding corrective vision devices like spectacles, which include an electronic device.

(i) **Section 110—Search Powers**[66]

Sub-clause (h) & (i) of the provisions state that to access a computer system or data storage device during a search, reasonable measures should be taken if the intangible material being searched for may be present in that system

65 Section 3, Search and Surveillance Act 2012 (2012 No. 24), Retrieved from: https://www.legislation.govt.nz/act/public/2012/0024/latest/DLM2136542.html?search=sw_096be8ed81df9dbe_electronic_25_se&p=1&sr=1.

66 Section 110, Search and Surveillance Act 2012 (2012 No. 24), Retrieved from: https://www.legislation.govt.nz/act/public/2012/0024/latest/DLM2136804.html?search=sw_096be8ed81df9dbe_computer_25_se&p=1&sr=2.

or device. If any accessed intangible material is relevant to the search or can be lawfully seized, it is allowed to make copies of that material using methods like previewing, cloning, or other forensic techniques, either before or after removing it for further examination.[67]

(j) **Section 130—Duty of persons with knowledge of computer system or other data storage devices or Internet site to assist access[68]**

The provision states that during a search of data in a computer system or data storage device, the person conducting the search can request a specified person[69] to provide access information and other reasonable assistance necessary for accessing the data. However, the specified person cannot be compelled to give information that may incriminate them. This restriction doesn't prevent the search personnel from requiring assistance to access data that may contain incriminating information. These provisions are subject to the regulations outlined in subpart 5 of this Part, addressing privilege and confidentiality.

(k) **Section 178—Offence of failing to carry out obligations in relation to computer system search[70]**

In accordance with Section 130 (as defined above), the provision penalizes a person failing to assist in exercising a search power when requested under Section 130(1). The offense is punishable by imprisonment for up to 3 months unless a reasonable excuse is provided.

---

67 Section 125(1)(l) & 125(1)(m) of the act provide similar powers to search a person, Retrieved from: https://www.legislation.govt.nz/act/public/2012/0024/latest/DLM2136824.html?search=sw_096be8ed81df9dbe_computer_25_se&p=1&sr=4.

68 Section 130, Search and Surveillance Act 2012 (2012 No. 24), Retrieved from: https://www.legislation.govt.nz/act/public/2012/0024/latest/DLM2136831.html?search=sw_096be8ed81df9dbe_computer_25_se&p=1&sr=5.

69 As defined under Section 130(5) a specified person includes a user with relevant knowledge of the system or device and a person managing an Internet site holding access information. A user refers to someone who owns, leases, possesses, or controls the system, device, or site, or has authorized access to data on an Internet site, including employees of such individuals or entities.

70 Section 178, Search and Surveillance Act 2012 (2012 No. 24), Retrieved from: https://www.legislation.govt.nz/act/public/2012/0024/latest/DLM2136894.html?search=sw_096be8ed81df9dbe_computer_25_se&p=1.

6. **Data Protection and Privacy Laws**
   New Zealand has data protection and privacy laws, such as the Privacy Act 2020[71], which can impact the admissibility of electronic evidence in cases involving breaches of privacy or data protection violations. To cite a few provisions the act that if the personal information requested by an individual is in a document, that information may be made available electronically or as a hard copy[72]. The act includes electronic records of personal information under the biometric information of a person[73]. The act lays various principles for Information Privacy under Section 22[74] of the act.

7.4 EVIDENCE LAWS IN AUSTRALIA

In Australia, the admissibility of electronic evidence is governed by various laws and legal principles.

1. **Evidence Act 1995**[75]
   The Evidence Act 1995 is the primary legislation governing the admissibility of evidence in Australia. It contains provisions that are relevant to electronic evidence, including definitions, admissibility criteria, and rules related to the use of documents. Key provisions of the act include:
   (a) **Section 71—Exception: Electronic Communications**[76]
   In electronic communications, there is an exception to the hearsay rule. This exception applies to representations in a document recording an electronic communication,

71 Privacy Act 2020 (2020 No. 31), Retrieved from: https://www.legislation.govt.nz/act/public/2020/0031/latest/LMS23223.html.

72 Section 56(1)(b), Privacy Act 2020 (2020 No 31), Retrieved from: https://www.legislation.govt.nz/act/public/2020/0031/latest/LMS23403.html?search=sw_096be8ed81de6f2f_electronic_25_se&p=1&sr=3.

73 Section 164, Privacy Act 2020 (2020 No. 31), Retrieved from: https://www.legislation.govt.nz/act/public/2020/0031/latest/LMS23610.html?search=sw_096be8ed81de6f2f_electronic_25_se&p=1&sr=4.

74 Section 22, Privacy Act 2020 (2020 No. 31), Retrieved from: https://www.legislation.govt.nz/act/public/2020/0031/latest/LMS23342.html?search=sw_096be8ed81de6f2f_individual_25_se&p=1&sr=7.

75 Evidence Act 1995 (No. 2, 1995), Retrieved from: https://www.legislation.gov.au/C2004A04858/latest/text.

76 Section 71, Evidence Act 1995 (No. 2, 1995), Retrieved from: https://www.austlii.edu.au/cgi-bin/viewdoc/au/legis/cth/consol_act/ea199580/s71.html.

specifically regarding the identity of the sender, the date and time of the communication, or the destination and addressee of the communication.

(b) **Section 161—Electronic Communications**[77]

The section establishes presumptions regarding electronic communications in Australia. If a document claims to contain a record of an electronic communication (excluding specific cases mentioned in Section 162[78]), it is presumed, unless evidence challenges it, that the communication was sent or made in the indicated form, by or on behalf of the represented person, on the specified day, time, and place, and was received at the mentioned destination. However, these presumptions do not apply in contract-related proceedings where all parties are involved in the contract, and the presumption conflicts with a contractual term.

2. **Electronic Transactions Act 1999**[79]

The Electronic Transactions Act 1999 deals with the legal recognition of electronic transactions and records in Australia. While not focused on admissibility, it lays the foundation for electronic records to be used in legal proceedings by Removing legal barriers to the use of electronic communications for transactions and recordkeeping, setting out rules for the legal validity of electronic signatures and documents, and addressing issues like time and place of dispatch/receipt of electronic communications.[80]

---

77 Section 161, Evidence Act 1995 (No. 2, 1995), Retrieved from: https://www.austlii.edu.au/cgi-bin/viewdoc/au/legis/cth/consol_act/ea199580/s161.html.

78 The section outlines presumptions related to messages transmitted through lettergrams or telegrams in Australia. If a document claims to contain a record of such a message, it is presumed, unless evidence challenges it, that the message was received by the addressee 24 hours after being delivered to a post office for transmission. However, this presumption does not apply in contract-related proceedings where all parties are involved in the contract, and the presumption conflicts with a term of the contract. Section 162, Evidence Act 1995 (No. 2, 1995), Retrieved from: https://www.austlii.edu.au/cgi-bin/viewdoc/au/legis/cth/consol_act/ea199580/s162.html.

79 Electronic Transactions Act 1999 (Act No. 162 of 199), Retrieved from: https://www.legislation.gov.au/C2004A00553/latest/text.

80 Electronic transactions legislation: An Australian perspective–core, Retrieved from: https://core.ac.uk/download/pdf/216910154.pdf.

3. **Privacy Laws and Data Protection**
   Privacy laws and data protection regulations in Australia, such as the Privacy Act 1988[81] and the Notifiable Data Breaches (NDB) scheme[82], may be relevant in cases involving electronic evidence, particularly when dealing with data breaches and privacy violations. The Privacy Act 1988 itself doesn't directly address the admissibility of evidence in court. However, its principles regarding data collection, use, disclosure, and storage can influence how courts assess the legality and ethical implications of using electronic evidence obtained through potentially privacy-invasive means[83]. The NDB scheme focuses on notifying individuals and the regulator about data breaches involving personal information. While breaches wouldn't automatically render evidence inadmissible, they could raise questions about data security and potentially lead to other legal consequences for the party responsible.[84]

LET'S RECALL

This chapter explores the legal frameworks and regulations governing the admissibility and handling of electronic evidence in four countries: The United States, the United Kingdom, New Zealand, and Australia.

- **United States**
  In the United States, electronic evidence laws, governed by federal and state rules, address admissibility, collection, and presentation in legal proceedings. Federal Rules of Evidence, including Rule 901 for authentication and Rule 902 for self-authentication, guide the process. Rule 803 outlines exceptions to the hearsay rule, ensuring reliability, and Rule 1001 defines terms relevant to electronic evidence. Landmark cases like Zubulake emphasize the duty to preserve electronic evidence, *United States v. Mitra* underscores

81 Privacy Act 1988 (No. 119, 1988), Retrieved from: https://www.legislation.gov.au/C2004A03712/2014-03-12/text.

82 The Notifiable Data Breaches (NDB) scheme, Retrieved from: https://www.oaic.gov.au/__data/assets/pdf_file/0011/5213/the-ndb-scheme-an-overview.pdf.

83 Privacy | Attorney-General's Department, Retrieved from: https://www.ag.gov.au/rights-and-protections/privacy.

84 Notifiable Data Breaches scheme | HLB Mann Judd, Retrieved from: https://hlb.com.au/notifiable-data-breaches-scheme-australia/.

the importance of a clear chain of custody, and decisions like *Riley v. California* highlight the need for proper legal procedures, contributing to the evolving landscape of electronic evidence in US courts.

- **United Kingdom**

  In the United Kingdom, the admissibility of electronic evidence is regulated by various laws and regulations. The Civil Evidence Act 1995 addresses admissibility in civil proceedings, with Section 5 emphasizing the reliability of the electronic document's creation process. The Criminal Justice Act 2003 introduces exceptions to the hearsay rule for specific electronic evidence, emphasizing admissibility conditions such as adequacy of computer systems and authentication. The Data Protection Act 2018 includes Schedule 22, focusing on the processing of personal data for law enforcement. The Civil Procedure Rules (CPR) and Criminal Procedure Rules (CrimPR) outline procedures for handling electronic evidence in civil and criminal cases, respectively. The Regulation of Investigatory Powers Act 2000 (RIPA) regulates electronic surveillance. Common law principles and case laws, including *R v. Bowden* (2005) and *Barclays Bank Plc v. Devonport Royal Dockyard Ltd* (2008), underscore the importance of reliability, integrity, and proper handling of electronic evidence in legal proceedings. The legal landscape continually evolves to address complexities in presenting electronic evidence, as demonstrated by cases such as *Serco Ltd v. Secretary of State for Defence* (2011), *Dawson-Damer & Ors v. Taylor Wessing LLP* (2017), and *Alibi v. CitySprint UK Ltd* (2019).

- **New Zealand**

  In New Zealand, the admissibility of electronic evidence is governed by several laws and legal principles. The Evidence Act 2006 is a primary legislation, defining "document" to include data held electronically and establishing criteria for admissibility. Section 106 outlines procedures for admitting video record evidence in criminal proceedings. Section 136 allows the verification of signatures through digital means. The Electronic Transactions Act 2002 (repealed by the Contract and Commercial Law Act 2017) and the Contract and Commercial Law Act 2017 provide a legal framework for electronic transactions, emphasizing clarity and equivalence with paper-based information. The Criminal Disclosure Act 2008 ensures proper disclosure of electronic evidence in criminal cases, while the Search and Surveillance Act 2012 regulates search and seizure powers, formalizing digital forensic procedures as law. Data protection and privacy laws, such as the Privacy Act 2020, impact the admissibility of electronic evidence, particularly in cases involving privacy breaches or data protection violations. The legal framework continually evolves, with cases like *R v. Bowden* (2005) and *Barclays Bank Plc v. Devonport Royal Dockyard Ltd* (2008) shaping principles around reliability, integrity, and proper handling of electronic evidence in New Zealand courts.

- **Australia**

  In Australia, the admissibility of electronic evidence is governed by key laws and principles. The Evidence Act 1995 is the primary legislation, featuring Section 71 which establishes a hearsay rule exception for electronic communications, specifically addressing representations in documents recording such communications. Section 161 of the act establishes presumptions related to electronic communications, unless challenged, regarding the form, sender's identity, timing, and destination. The Electronic Transactions Act 1999 sets the foundation for electronic transactions and records in legal proceedings, addressing electronic signatures and the legal validity of electronic documents. Privacy laws, including the Privacy Act 1988 and the Notifiable Data Breaches scheme, are relevant in cases involving electronic evidence, influencing the assessment of legality and ethics, especially in instances of data breaches and privacy violations. While these laws don't directly address evidence admissibility, they impact how courts consider electronic evidence obtained through potentially privacy-invasive means.

Knowledge Check

A. CHOOSE THE CORRECT OPTION

1. In the context of electronic evidence, what is the importance of a well-maintained chain of custody?
   (a) Ensuring encryption
   (b) Demonstrating authenticity and relevance
   (c) Privacy protection
   (d) Admissibility of evidence

2. What case emphasized the need for the proper handling and authentication of electronic records and communications in criminal proceedings?
   (a) *R* v. *Bowden* (2005)
   (b) *R* v. *Tisi* (2006)
   (c) *Dawson-Damer & Ors v. Taylor Wessing LLP* (2017)
   (d) *Alibi* v. *CitySprint UK Ltd* (2019)
3. In New Zealand, which legislation contains provisions related to the admissibility of electronic evidence, including definitions of documents and admissibility criteria?
   (a) Evidence Act 2006
   (b) Electronic Transactions Act 2002
   (c) Criminal Disclosure Act 2008
   (d) Search and Surveillance Act 2012
4. Under Rule 901 of the Federal Rules of Evidence, what does authentication of evidence involve?
   (a) Admissibility of hearsay evidence
   (b) Proving the evidence is what it purports to be
   (c) Exceptions to the hearsay rule
   (d) Definitions relevant to evidence in modern legal contexts
5. Which Australian legislation deals with the legal recognition of electronic transactions and records, laying the foundation for electronic records to be used in legal proceedings?
   (a) Evidence Act 1995
   (b) Electronic Transactions Act 1999
   (c) Privacy Act 1988
   (d) Criminal Procedure Act 1986 (NSW)

Answer Key

1. (b) Demonstrating authenticity and relevance
2. (a) *R* v. *Bowden* (2005)
3. (a) Evidence Act 2006
4. (b) Proving the evidence is what it purports to be
5. (b) Electronic Transactions Act 1999

B. ANSWER THE FOLLOWING

1. Discuss and Elaborate on Australian laws on Electronic Evidence.
2. Examine the critical role of a secure chain of custody in ensuring the admissibility of electronic evidence. Highlight how meticulous

chain of custody management is essential for establishing the authenticity and reliability of electronic evidence. Provide references to relevant provisions and legal precedents across different jurisdictions.

3. How does the Search and Surveillance Act 2012 lay the foundation for the use of electronic records in legal proceedings and set out procedures for digital forensic investigations in New Zealand?
4. Provide an overview of the Federal Rules of Evidence in the United States. How do these rules govern the admissibility of electronic evidence in legal proceedings? Choose one of the listed US cases (e.g., *Riley v. California*) and explain its significance in the context of electronic evidence.
5. How do Sections 115 and 116 of the Criminal Justice Act 2003 address the admissibility of computer-generated evidence and electronic communications in criminal trials?

CHAPTER 8

E-Witness
Digital Evidence Alibis and Testimony of Bits & Bytes

A Glimpse into the Chapter

- The Testimony of Bits & Bytes—Key attributes and importance of electronic evidence as a witness in legal proceedings
- Digital Alibi—Illustrations as to how electronic evidence can be used as an alibi
- Challenges with respect to E-Witness
- Judicial Pronouncements

Digital, Internet, online, etc., a decade back, these were just terms; however, today, they are "prefix" to almost everything on this planet: Digital India, Internet Banking, Online Communication, e-Commerce, e-business & Online Classes. This digital era has dramatically changed the landscape of legal procedures, fundamentally altering how evidence is brought and analyzed within the sacred halls of the courtroom. Electronic evidence is the foundation of this revolutionary process; it is a broad category that includes a multitude of digital artefacts, such as files, messages, records, and other data. It has swiftly risen to prominence, becoming an integral component in the machinery of justice. Its role has diversified with respect to a witness and an alibi, functioning as a silent yet compelling observer of past events and an unwavering source of corroborative support for legal contentions. Doing so brings a hitherto unseen level of accuracy and dependability to court procedures.

8.1 E-WITNESS: AN INTRODUCTION

The digital revolution, including the legal system, has impacted every facet of modern society.[1] With the development of technology, the idea of an e-Witness has surfaced, upending preconceived ideas about testifying in court. Electronic witnesses use digital tools and technologies to record and authenticate events, unlike traditional witnesses who testify in person. Electronic witnessing encompasses a variety of forms, ranging from audiovisual recordings to digital signatures and electronic notarization. The definition of e-Witness extends beyond traditional witnesses and includes any electronic means used to capture, record, or verify events relevant to legal proceedings. Common types of e-Witness include video recordings, digital signatures, timestamped documents, and electronic notarization.[2] Legal admissibility is a crucial factor to take into account when integrating e-Witness. In evaluating electronic evidence, courts have to ensure that it is authentic and dependable to the same extent as traditional testimony. The potential for digital record manipulation, hacking, or tampering presents difficulties and doubts e-Witness credibility in court.[3]

Using electronic witnesses raises ethical questions about security, consent, and privacy. The consent of individuals to be recorded, the safe storage of digital evidence, and the possibility of technology misuse are all concerns that need to be carefully considered when using electronic devices that gather and store data. Ethical rules and regulations are essential to protect the rights and interests of parties involved in legal proceedings.[4] Despite the challenges, e-Witness offers several potential benefits to the legal system. Electronic witnesses can provide more accurate and reliable accounts of events, reduce the reliance on human memory, and streamline legal processes through the use of advanced technologies.[5] Additionally,

---

1 Gill, N., & Hynes, J. (2021). Courtwatching: Visibility, publicness, witnessing, and embodiment in legal activism. Area, 53(4), 569–576.

2 Awan, N. (2021). Digital witnessing and the erasure of the racialized subject. *Journal of Visual Culture*, 20(3), 506–521.

3 Kassin, S. M., Ellsworth, P. C., & Smith, V. L. (1989). The" general acceptance" of psychological research on eyewitness testimony: A survey of the experts. *American Psychologist*, 44(8), 1089. Also see: Doak, J., & McGourlay, C. (2009). Criminal evidence in context. Routledge.

4 Chabon, S., Morris, J., & Lemoncello, R. (2011, November). Ethical deliberation: A foundation for evidence-based practice. In *Seminars in speech and language* (Vol. 32, No. 04, pp. 298–308). © Thieme Medical Publishers.

5 Duranti, L., Eastwood, T., Eastwood, T., & MacNeil, H. (2002). *Preservation of the integrity of electronic records*. Springer Science & Business Media.

the use of e-Witness may enhance accessibility to justice by facilitating remote testimony and reducing the need for physical presence in court.[6]

8.2 THE TESTIMONY OF BITS AND BYTES

In the ever-evolving digital technology landscape, electronic evidence often takes on the role of a silent yet compelling witness. It can be regarded as a witness because it provides a factual account of events that transpired, and its credibility is often on par with traditional human witnesses. Electronic evidence serves as a witness in the sense that it offers factual and objective information about a particular incident or set of circumstances. This information can be used to establish the occurrence of events, corroborate or challenge testimonies, and support legal claims.[7] Electronic evidence as a witness refers to using digital records, data, or other electronic information in legal proceedings to provide first-hand accounts of events, actions, or communications. It can include various types of digital records, such as emails, text messages, surveillance footage, and GPS data, which record events and actions.[8]

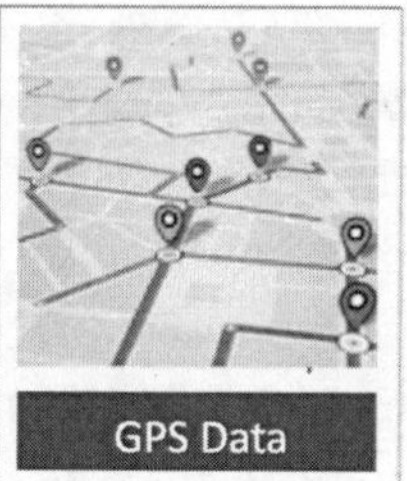

Figure 8.1 Pictorial Representation of Digital Records.

Unlike human witnesses, electronic evidence is typically unbiased, unemotional, consistent, and persistent, making it a reliable source of information in court.[9] Key attributes of electronic evidence and its importance as a witness in legal proceedings are mentioned below:

---

6 Implications of the electronic witnessing provisions retrieved from: https://www.lawsociety.com.au/sites/default/files/2021-12/Implications%20of%20Electronic%20Witnessing%20Provisions.pdf, visited on 23 January 2023.

7 Dekeyser, H. (2006). Authenticity in bits and bytes. Also see: Highland, H. J. (1997). Historical bits & bytes. *Computers & Security*, 16(5), 387-411.

8 Barkett, J. M. (2004). Bytes, Bits and Bucks: Cost Shifting and Sanctions in e-Discovery. *Def. Counsel J.*, 71, 334.

9 McCrystal, J. L., & Maschari, A. B. (1979). Will Electronic Technology Take the Witness Stand. *U. Tol. L. Rev.*, 11, 239.

1. **Immediacy and Accuracy:** One of the most prominent qualities of electronic evidence is its immediacy and accuracy. Digital records are generated in real-time, whether text messages, emails, surveillance footage, social media posts, GPS data, or any other form of electronic documentation. This real-time evidence generation lends a high degree of accuracy to the information. For instance, the timestamp on an email or a GPS location can precisely establish when and where an event occurred. In contrast to human witnesses whose recollections may be flawed by the passage of time, electronic evidence remains steadfast and reliable.[10]
2. **Unbiased and Unemotional:** Electronic evidence lacks emotions, biases, or distortions that often influence human witnesses.[11] It presents facts as they are, without personal interpretations or emotional filters. This absence of emotional bias makes it a more reliable source of information in many instances.
3. **Consistency:** Digital records are consistent and do not change their statements over time. Unlike human witnesses whose memories can fade, alter, or be influenced, electronic evidence remains consistent and authentic to its original form. This consistency is a key attribute that contributes to its reliability.[12]
4. **Persistence:** Once created, electronic evidence persists over time, making it available for review and analysis long after an event has transpired. This persistence is of critical importance in the realm of investigations and legal proceedings. While human memories may fade or become unreliable, electronic evidence remains intact and can be referenced whenever necessary.
5. **Corroboration:** Electronic evidence often complements and corroborates testimony from human witnesses. When the evidence aligns with individuals' accounts, it reinforces their statements' credibility. In this way, electronic evidence serves as a witness in its own right and as a supporter of human witnesses, enhancing the overall reliability of the evidence presented in court.

---

10 Widdison, R. (1997). Electronic Law Practice: an exercise in legal futurology. *Mod. L. Rev.,* 60, 143.

11 Wiener, R. L., Bornstein, B. H., & Voss, A. (2006). Emotion and the law: A framework for inquiry. *Law and human Behavior,* 30, 231–248.

12 Duranti, L., & Rogers, C. (2012). Trust in digital records: An increasingly cloudy legal area. *Computer Law & Security Review,* 28(5), 522–531.

6. **Examination by Computer:** If the computer was examined within a few days of its use, a computer expert witness would find strong evidence that it was used on a specific date and time. Continuous use of the computer reduces the likelihood of finding evidence of use. Evidence of computer use can last for weeks or months, and an expert witness will uncover any remaining evidence.

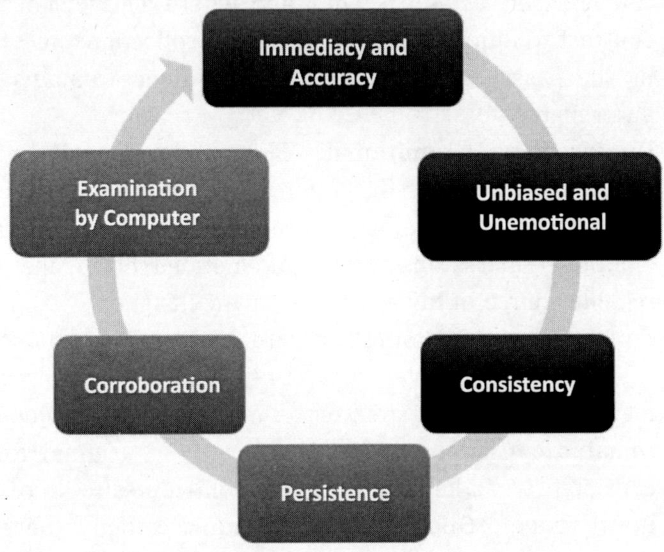

Figure 8.2 Attributes of Electronic Evidence as Witness.

8.3 DIGITAL ALIBI

An alibi is a legal defence that provides a verifiable account of an individual's whereabouts and actions during a specific time frame, suggesting that they could not have committed the alleged offence.[13] In criminal cases, where the accused's presence at a different location can potentially disprove their involvement in the crime, electronic evidence often plays a pivotal role. One of the most compelling uses of electronic evidence is as an alibi. Listed below are a few examples that depict how electronic evidence can be used as an alibi:

1. **Times and Dates:** When the computer is being used, it records dates and times in various places. If an individual is ever accused

13 Castiglione, A., Cattaneo, G., De Maio, G., De Santis, A., Costabile, G., & Epifani, M. (2012, July). The forensic analysis of a false digital alibi. *In 2012 Sixth International Conference on Innovative Mobile and Internet Services in Ubiquitous Computing* (pp. 114–121). IEEE.

of being somewhere else, he/she could almost certainly use evidence from his/her computer to prove that he/she was at his/her desk, using his/her computer. Examining the computer and its circumstances will reveal the dependability of the Accessed Date on a computer.

2. **Event Logs:** The Windows operating system logs messages about events in the Event Log.[14] Messages recording the date and time a computer is turned on and off are notable examples of these event log messages. Some software is constantly running on the computer to perform housekeeping tasks. Even when the computer appears to be doing nothing, it generates event logs.
3. **Internet Exploration:** Some software programs include features and functions for storing dates and times. Web browsers are the examples of software that store dates and times.[15] Everything displayed on the screen while browsing the web is also saved on the computer. When you view a single web page, several files are created and saved on your computer. Each of these files will have a creation date. One could spend an hour surfing the web and clicking from page to page. One or more files will be saved for each page that one views. A subsequent examination of the Creation Dates for these files will reveal the period of web browsing. Web browsers include a history feature. It can display a list of all the websites an individual visited previous day and in the previous thirty days. This History is saved on the computer and can be viewed.
4. **Other Software:** Numerous software programs save information from previous uses. Some software stores revision and version information in data files so that we can go back to a previous version of a document. The computer must be examined to determine the files that have been saved and the software that was used to create these files. This investigation may uncover unexpected sources of evidence regarding computer use.

---

14 Talebi, J., Dehghantanha, A., & Mahmoud, R. (2015). Introducing and analysis of the Windows 8 event log for forensic purposes. In *Computational Forensics: 5th International Workshop, IWCF 2012,* Tsukuba, Japan, November 11, 2012 and *6th International Workshop, IWCF 2014,* Stockholm, Sweden, August 24, 2014, Revised Selected Papers (pp. 145-162). Springer International Publishing.

15 Graham Dilloway CITP MBCS.
Computer Expert Witness, Retrieved from: https://www.dilloway.co.uk/computer-as-alibi-witness.html, visited on 23 January 2024.

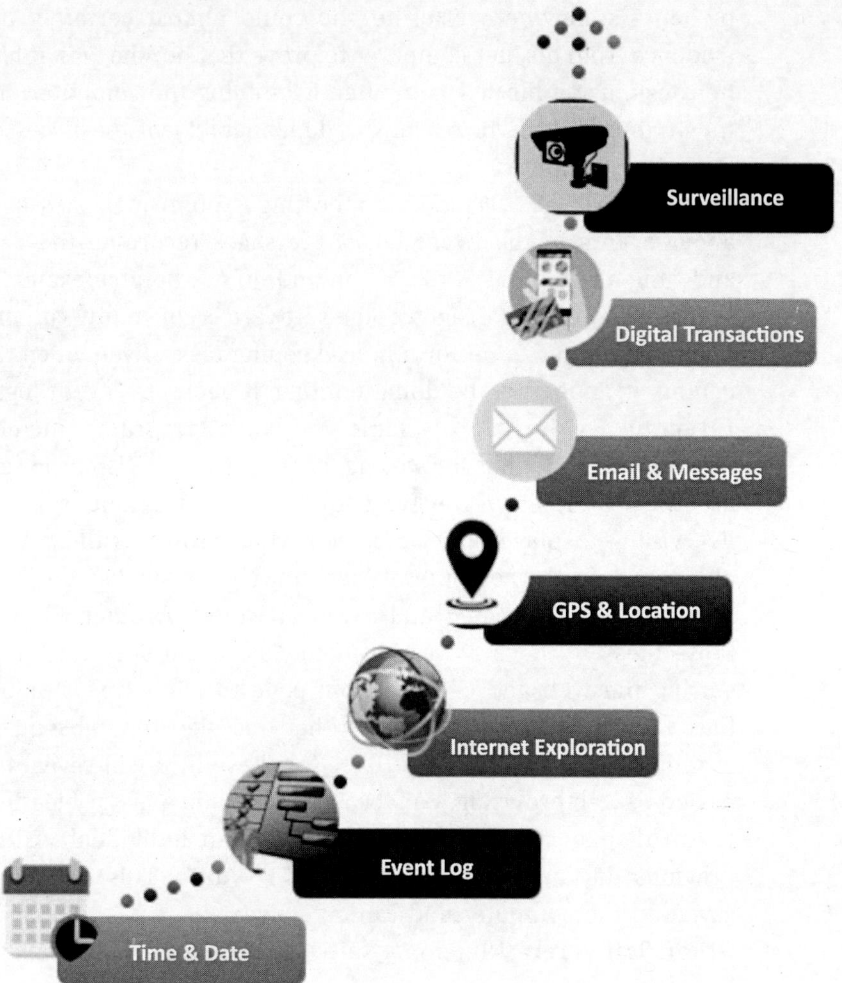

Figure 8.3 Examples of Electronic Evidence as Alibi.

5. **GPS and Location Data:** Mobile phones and various other digital devices are equipped with GPS capabilities. These devices record the location of the user in real time. GPS data can be invaluable as an alibi, as it can provide a detailed record of an individual's movements. If these records show that the person was at a different location at the time of the alleged incident, it can serve as a strong and compelling alibi. For example, GPS data can indicate that the accused was miles away from the alleged crime scene when it occurred, making their involvement highly unlikely.
6. **Timestamped Messages and Emails:** Many forms of electronic communication, such as text messages, emails, and chat logs, include timestamps that specify the exact moment when a message

was sent or received. Timestamps are often utilized as part of an alibi to prove that the accused was engaging in conversations or sending correspondence from a specific location at a particular time. When presented as evidence, these timestamped messages can establish that the person was elsewhere when the crime took place.

7. **Digital Records of Transactions:** In many cases, digital records of financial transactions can be used to establish an alibi. Credit card transactions, ATM withdrawals, and digital payment records create a verifiable trail that can place an individual at a specific location at a particular time. These records can serve as compelling evidence, demonstrating that the person was far from the scene of the alleged offense when it occurred.
8. **Surveillance Footage:** In locations equipped with security cameras, video recordings can provide visual evidence of a person's presence at a specific location. Surveillance footage is often used in legal proceedings to serve as an alibi, confirming an individual's location at a particular time. The presence of a person in surveillance footage at a different location can be a compelling defense.

8.3.1 Using Electronic Evidence to Establish Alibi: Illustrations

Scenario 1: The defendant's mobile phone records, including call logs and location data, can also serve as electronic evidence. They show that the defendant's phone was connected to a cell tower near the second store during the time of the alleged theft, supporting the alibi that the defendant was not present at the scene of the crime.

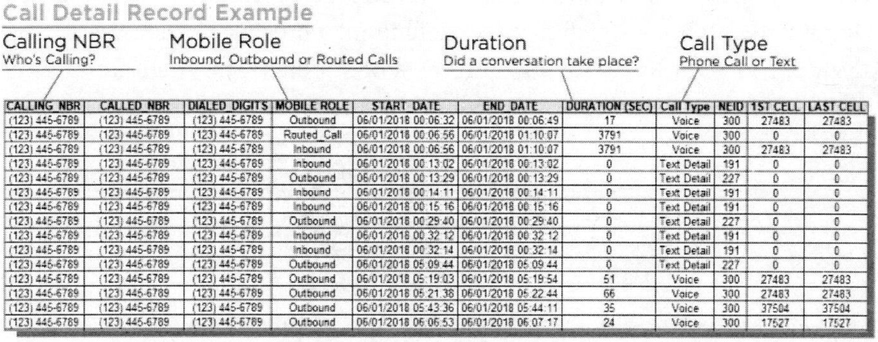

Call Detail Record Example

Calling NBR — Who's Calling?
Mobile Role — Inbound, Outbound or Routed Calls
Duration — Did a conversation take place?
Call Type — Phone Call or Text

CALLING_NBR	CALLED_NBR	DIALED_DIGITS	MOBILE ROLE	START_DATE	END_DATE	DURATION (SEC)	Call Type	NEID	1ST CELL	LAST CELL
(123) 445-6789	(123) 445-6789	(123) 445-6789	Outbound	06/01/2018 00:06:32	06/01/2018 00:06:49	17	Voice	300	27483	27483
(123) 445-6789	(123) 445-6789	(123) 445-6789	Routed_Call	06/01/2018 00:06:56	06/01/2018 01:10:07	3791	Voice	300	0	0
(123) 445-6789	(123) 445-6789	(123) 445-6789	Inbound	06/01/2018 00:06:56	06/01/2018 01:10:07	3791	Voice	300	27483	27483
(123) 445-6789	(123) 445-6789	(123) 445-6789	Inbound	06/01/2018 00:13:02	06/01/2018 00:13:02	0	Text Detail	191	0	0
(123) 445-6789	(123) 445-6789	(123) 445-6789	Outbound	06/01/2018 00:13:29	06/01/2018 00:13:29	0	Text Detail	227	0	0
(123) 445-6789	(123) 445-6789	(123) 445-6789	Inbound	06/01/2018 00:14:11	06/01/2018 00:14:11	0	Text Detail	191	0	0
(123) 445-6789	(123) 445-6789	(123) 445-6789	Inbound	06/01/2018 00:15:16	06/01/2018 00:15:16	0	Text Detail	191	0	0
(123) 445-6789	(123) 445-6789	(123) 445-6789	Outbound	06/01/2018 00:29:40	06/01/2018 00:29:40	0	Text Detail	227	0	0
(123) 445-6789	(123) 445-6789	(123) 445-6789	Inbound	06/01/2018 00:32:12	06/01/2018 00:32:12	0	Text Detail	191	0	0
(123) 445-6789	(123) 445-6789	(123) 445-6789	Inbound	06/01/2018 00:32:14	06/01/2018 00:32:14	0	Text Detail	191	0	0
(123) 445-6789	(123) 445-6789	(123) 445-6789	Outbound	06/01/2018 05:09:44	06/01/2018 05:09:44	0	Text Detail	227	0	0
(123) 445-6789	(123) 445-6789	(123) 445-6789	Outbound	06/01/2018 05:19:03	06/01/2018 05:19:54	51	Voice	300	27483	27483
(123) 445-6789	(123) 445-6789	(123) 445-6789	Outbound	06/01/2018 05:21:38	06/01/2018 05:22:44	66	Voice	300	27483	27483
(123) 445-6789	(123) 445-6789	(123) 445-6789	Outbound	06/01/2018 05:43:36	06/01/2018 05:44:11	35	Voice	300	37504	37504
(123) 445-6789	(123) 445-6789	(123) 445-6789	Outbound	06/01/2018 06:06:53	06/01/2018 06:07:17	24	Voice	300	17527	17527

Figure 8.4 Call Record Details Example.[16]

16 Image Source: https://www.linkedin.com/pulse/call-detail-records-super-phone-bill-lars-daniel.

Scenario 2: The defendant paid for items at another store shortly after the alleged theft. Electronic evidence in the form of timestamped receipts from the second store indicates that the defendant was at a different location during the time of the theft.

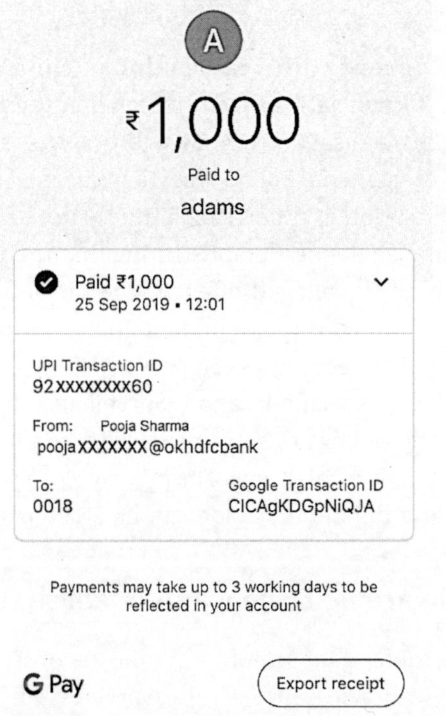

Figure 8.5 Screenshot of Payment Receipt.[17]

Scenario 3: The store has security cameras that capture the incident. The surveillance footage, considered electronic evidence, becomes a key witness in this case. It shows the defendant entering the store, moving around, and eventually leaving with the stolen items. The video timestamp provides an accurate account of when the theft occurred.

Figure 8.6 Screenshot of a CCTV Timestamp.[18]

17 Image Source: https://support.coindcx.com/articles/dcx-insta/what-is-reference transaction-id/6381b9b62418a07148a9356d.

18 Image Source: https://www.forensicfocus.com/articles/when-did-it-happen-dealing-with-timestamps-in-amped-five/.

Scenario 4: In the case of harassment or threats, text message records serve as electronic witnesses. These records include the content of threatening messages, timestamps, and the phone numbers involved. They provide an unemotional and consistent account of the communications, helping to establish the facts of the case.

Scenario 5: In a business dispute, email correspondence between parties can serve as electronic witnesses. The emails may include agreements, discussions, and commitments. Their timestamps and content offer an accurate account of the negotiations and commitments made by the parties involved.

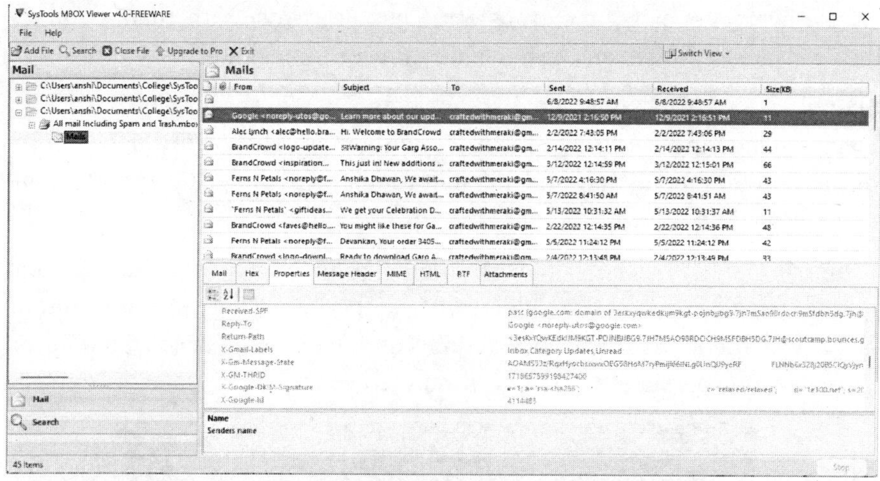

Figure 8.7 Artefacts extracted from an email using Systools MBOX Viewer Tool.

Scenario 6: In a defamation case, social media posts can act as electronic witnesses. The posts may contain defamatory statements or evidence of false claims. The timestamps on these posts provide immediate and unemotional records of when the statements were made and by whom.

Scenario 7: In a personal injury case involving a car accident, the GPS data from the vehicles involved serves as electronic witnesses. It provides a detailed and accurate account of the vehicles' locations and speeds leading up to the accident. This information is critical in reconstructing the events and determining liability.

	T	U	V	W	
1	Value.placeVisit.location.latitudeE7	Value.placeVisit.location.longitudeE7	Value.placeVisit.location.placeId	Value.placeVisit.location.address	Value.placeVisit.loca
2					
3					
4					
5	285960150	770489000	ChIJsSWnntAaDTkRlTldmexi9cQ	Sector 12, Dwarka, New Delhi, South West Delhi, Delhi 110075, India	Surya Tower
6	285960150	770489000	ChIJsSWnntAaDTkRlTldmexi9cQ	Sector 12, Dwarka, New Delhi, South West Delhi, Delhi 110075, India	Surya Tower
7	285960150	770489000	ChIJsSWnntAaDTkRlTldmexi9cQ	Sector 12, Dwarka, New Delhi, South West Delhi, Delhi 110075, India	Surya Tower
8	285960150	770489000	ChIJsSWnntAaDTkRlTldmexi9cQ	Sector 12, Dwarka, New Delhi, South West Delhi, Delhi 110075, India	Surya Tower
9	285960150	770489000	ChIJsSWnntAaDTkRlTldmexi9cQ	Sector 12, Dwarka, New Delhi, South West Delhi, Delhi 110075, India	Surya Tower
10	285960150	770489000	ChIJsSWnntAaDTkRlTldmexi9cQ	Sector 12, Dwarka, New Delhi, South West Delhi, Delhi 110075, India	Surya Tower
11	285960150	770489000	ChIJsSWnntAaDTkRlTldmexi9cQ	Sector 12, Dwarka, New Delhi, South West Delhi, Delhi 110075, India	Surya Tower
12	285960150	770489000	ChIJsSWnntAaDTkRlTldmexi9cQ	Sector 12, Dwarka, New Delhi, South West Delhi, Delhi 110075, India	Surya Tower
13	285960150	770489000	ChIJsSWnntAaDTkRlTldmexi9cQ	Sector 12, Dwarka, New Delhi, South West Delhi, Delhi 110075, India	Surya Tower
14					
15					

Figure 8.8 Google Maps data extracted from a user account. It holds a number of columns including location address & name, timestamp of the location, longitude & latitude, etc.

Scenario 8: In cases involving financial disputes, bank transaction records act as electronic witnesses. They include detailed information about withdrawals, deposits, and transfers. These records, with their accuracy and consistency, offer an objective account of financial transactions.

Scenario 9: In a contract dispute, video conference recordings of a business meeting serve as electronic witnesses. The recordings capture discussions, agreements, and presentations, providing an accurate and unemotional account of the meeting's proceedings.

Scenario 10: In cybercrimes or online fraud cases, website logs and server records can act as electronic witnesses. They provide details about website access, user interactions, and activities. These logs offer an objective and persistent account of online events.

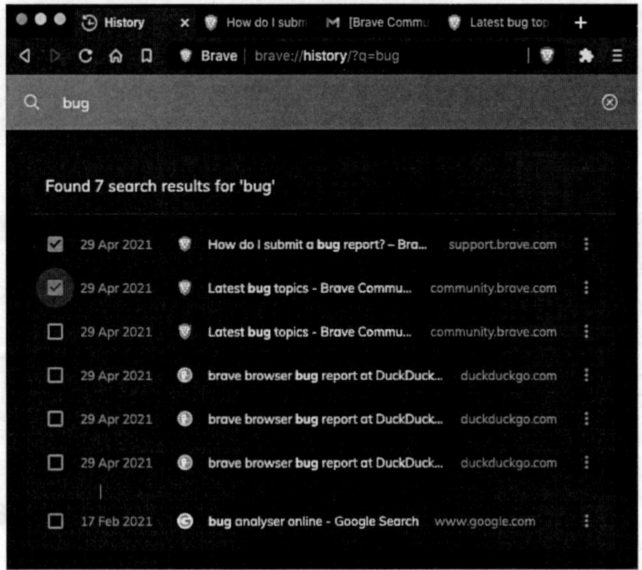

Figure 8.9 Screenshot of a web history.[19]

19 Image Source: https://community.brave.com/t/a-bug-in-ui-in-browser-history-page/240494.

Scenario 11: In aviation accidents, flight data recorders (commonly known as "black boxes") serve as electronic witnesses. They capture data about the aircraft's performance, altitude, speed, and other crucial information. This electronic evidence is indispensable in accident investigations.

Scenario 12: Every file on a computer has timestamps associated with it:

- When a file is given a name and saved to the computer, the Created Date is saved. This indicates when the file was initially saved.
- When the content of a file is changed, the Modified Date is updated. This date reflects the most recent modification made to the file.
- When a file is accessed for reading or updating, the Accessed Date is changed. If the last use of the file was to change its content, the Modified and Accessed Dates will be the same.
- After finishing typing, an hour was spent reading through the report. The Accessed Date would be one hour later than the Modified Date. Recent versions of Windows do not use the Accessed Date by default.

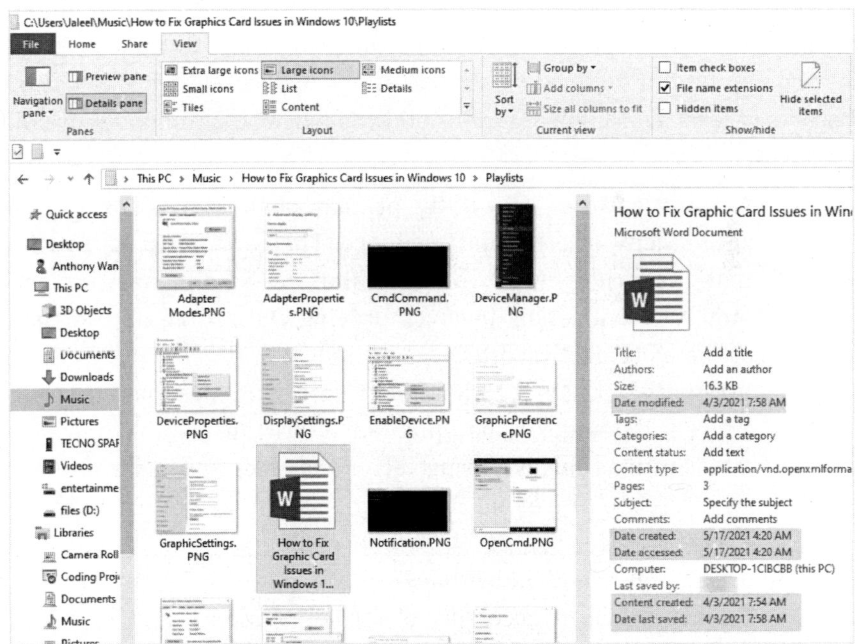

Figure 8.10 Example of a Timestamps/Metadata available in a file.[20]

20 Image Source: https://www.webnots.com/how to change-date-attributes-of-files-and-folders-in-windows-10/.

In each of these examples, electronic evidence serves as a reliable and objective witness by providing immediate, accurate, unbiased, consistent, and persistent records of events or communications. It plays a critical role in helping courts and legal professionals establish the facts and arrive at just outcomes in various legal proceedings.

Electronic evidence is not limited to criminal cases alone. In civil cases, electronic evidence can be just as vital in establishing alibis in various contexts. In contract disputes, employment-related matters, or personal injury cases, digital records such as emails, text messages, and transaction logs can be used to corroborate an individual's or business's claims. These records serve as verifiable evidence to demonstrate that a person or entity acted according to their legal obligations or commitments.

8.4 CHALLENGES WITH RESPECT TO E-WITNESS

It is helpful to consider how new technology will alter recurring components of investigations. There will inevitably be more alibis that rely on digital evidence as people spend more time using mobile devices, computers, and networks. Although there are numerous ways that information stored on a computer can be used to determine when a computer was used, it presents several legal considerations and challenges that must be carefully navigated:[21]

1. **Authentication:** One of the primary challenges is ensuring the authenticity of electronic evidence. The court must be satisfied that the evidence is what it claims to be and has not been tampered with. This involves proving the source and integrity of the electronic records.
2. **Integrity:** Ensuring that electronic evidence has not been altered, deleted, or manipulated is crucial. Establishing data integrity may involve using techniques like hashing and digital signatures to prove that the evidence has remained unchanged.[22]
3. **Privacy:** The use of electronic evidence must comply with privacy laws and individuals' rights. There is often a delicate balance between accessing electronic evidence and respecting

21 Samanta, P., & Jain, S. (2018, October). E-Witness: Preserve and prove forensic soundness of digital evidence. In *Proceedings of the 24th Annual International Conference on Mobile Computing and Networking* (pp. 832–834).

22 Forensic Examination of Digital Evidence: A Guide for Law Enforcement, retrieved from: https://www.ojp.gov/pdffiles1/nij/199408.pdf, visited on 23 January 2024.

privacy rights, particularly in cases involving personal or sensitive data.

4. **Chain of Custody:** Maintaining a clear and documented chain of custody is crucial. The chain of custody details the handling and storage of electronic evidence from its collection to its presentation in court. Any gaps or irregularities in the chain can raise doubts about the integrity of the evidence.
5. **Hearsay:** Electronic evidence may sometimes fall under the category of hearsay, where an out-of-court statement is introduced for the truth of the matter asserted. Traditional hearsay rules may present challenges in admitting electronic evidence, but exceptions to the hearsay rule can apply.
6. **Preservation:** Electronic evidence can be volatile. Efforts must be made to preserve the evidence in its original state. Failure to do so may lead to challenges related to spoliation, which involves the destruction of evidence.

Figure 8.11 Challenges with Electronic Evidence.

7. **Expertise**
   (a) **Expert Witnesses:** In many cases, experts in digital forensics or cybersecurity may be required to authenticate and explain electronic evidence. The court must decide whether such experts are qualified to testify.
   (b) **Technical Expertise**: Judges and attorneys may lack the technical expertise to fully understand complex electronic evidence, making it challenging to assess the evidence's validity and relevance.
8. **Admissibility and Evidentiary Value**
   (a) **Relevance:** The evidence must be relevant to the case at hand. Courts need to determine whether the electronic evidence is directly related to the issues being litigated.
   (b) **Character Evidence:** Electronic evidence may not be admissible if it is offered to prove a person's character and propensity to act in a certain way. The court must assess the evidence's purpose.
   (c) **Best Evidence Rule:** The best evidence rule requires that the original electronic evidence be presented if available, rather than copies or summaries.
   (d) **Evidentiary Value:** Courts must assess the probative value of electronic evidence. While it may be highly accurate and immediate, it is not immune to challenges regarding its reliability, context, and significance in the case.
9. **Jurisdiction:** Electronic evidence collected from servers or devices located in different jurisdictions may involve complex legal questions, including which laws apply and how evidence can be obtained. In cases involving electronic evidence from foreign sources, the legal process can be time-consuming and complex due to policies that govern the exchange of evidence between countries.
10. **Cost:** Electronic evidence can be costly to retrieve, especially when it involves specialized forensic services or the recovery of data from damaged or encrypted devices. This cost may affect the ability to access and present evidence.

8.5 JUDICIAL PRONOUNCEMENTS

Anand s/o Shivaji Ghodale v. State of Maharashtra: Bombay High Court[23]

In the recent case of Anand s/o Shivaji Ghodale v. State of Maharashtra, the Bombay High Court emphasized the importance of introducing the defense of alibi at an early stage of the trial, even during the framing of charges. The court ruled that there is no legal requirement for the defense of an alibi to be presented only during the stage of defense evidence. The petitioner in this case was able to provide electronic evidence in the form of CCTV footage that supported his alibi, demonstrating that he was not present at the scene of the alleged incident. The court highlighted that Section 319 of the Criminal Procedure Code should not be applied mechanically, and substantial evidence is required before summoning an individual to face trial. This judgment underscores the need for a thorough assessment of evidence and legal principles when deciding to summon individuals for trial.

Key relevant points noted from the case:

1. **CCTV Footage as Alibi:** The petitioner's defense of alibi was substantiated by the Chemical Analysis report of the bank's CCTV footage, which clearly showed the petitioner's presence at his workplace (a bank) at the time of the alleged offense. This electronic evidence played a pivotal role in supporting the alibi.
2. **Importance of Electronic Evidence:** The court highlighted the importance of electronic evidence, such as CCTV footage, as an admissible and credible source of information in legal proceedings. In this case, the petitioner's alibi was strongly supported by electronic evidence, further underscoring its significance in modern legal practice.

Global Scenario

Listed below are a few notable foreign cases that involve the use of electronic evidence and alibi defenses. These cases demonstrate how electronic evidence has been used in conjunction with alibi defenses in various jurisdictions:

1. **United States—State of Maryland v. Adnan Syed (2000)**
   Adnan Syed was convicted of murder, and his defense presented cell phone records as electronic evidence to support his alibi. The

23 Criminal Revision Application No. 296 of 2022, decided on January 23, 2023.

records showed that his cell phone was likely not near the crime scene at the time of the murder.

2. **United Kingdom—R v. Riat (2009)**
   In this case, the defendant was charged with a murder that occurred on a London street. The defense presented evidence from London's extensive network of CCTV cameras, showing that the defendant was far away from the crime scene when the murder occurred.
3. **Canada—R v. Oland (2015)**
   This high-profile Canadian case involved the murder of a wealthy businessman. The defendant's legal team used electronic evidence from the victim's cell phone to establish an alibi. The data indicated that the victim's phone was used after the estimated time of death, casting doubt on the defendant's guilt.
4. **Australia—R v. Wood (2012)**
   In this Australian case, the accused was charged with murder. The defense presented electronic evidence from a GPS tracking device attached to the accused's ankle. The data was used to support the defendant's claim that he was at a different location when the murder occurred.
5. **South Africa—S v. Phadima (2014)**
   This South African case involved a murder trial where the accused presented evidence from the cell phone records of both the victim and the accused. The electronic evidence was used to establish the accused's alibi, suggesting that the victim had a different location and was in contact with someone else at the time of the crime.

8.6 CONCLUSION

The emergence of e-Witness marks a significant shift in the landscape of legal proceedings. As technology advances, the legal system must adapt to incorporate electronic evidence while addressing the associated challenges and ethical considerations. A carefully crafted regulatory framework is imperative to ensure the reliability and admissibility of e-Witness testimony, ultimately enhancing the efficiency and accessibility of the justice system. Navigating legal considerations and challenges in the use of electronic evidence as a witness and alibi requires a combination of legal expertise and technological know-how. Legal professionals, expert witnesses, and the judiciary must work together to ensure that electronic evidence is handled, presented, and evaluated to uphold the principles of fairness, authenticity, and justice in the legal system. As technology advances, the legal landscape for electronic evidence will evolve, requiring ongoing adaptation and expertise.

It is essential to recognize that while electronic evidence can be a potent tool for establishing an alibi or presenting a witness account of events, its admissibility and authenticity must be rigorously assessed. The legal standards and guidelines surrounding the use of electronic evidence in court are evolving in tandem with technological advancements. It is imperative to ensure that the chain of custody is properly documented and maintained, that data integrity is preserved, and that all legal standards are adhered to guarantee electronic evidence's reliability and admissibility.

LET'S RECALL

The digital era has significantly transformed legal procedures, fundamentally changing how evidence is presented and analyzed within the courtroom. Electronic evidence plays a central role in this shift.

- Electronic evidence, often referred to as e-witness, is introduced as a crucial component in legal proceedings. It serves as a factual, unbiased, and consistent witness that can provide immediate, real-time accounts of events.
- The key attributes and the importance of electronic evidence as a witness are:
    - *Immediacy and Accuracy:* Electronic evidence is generated in real time, providing high accuracy.
    - *Unbiased and Unemotional:* It lacks emotional bias, making it a reliable source of information.
    - *Consistency:* Electronic records remain consistent over time.
    - *Persistence:* Once created, electronic evidence persists over time, making it available for review.
    - *Corroboration:* Electronic evidence often complements and corroborates testimony from human witnesses, enhancing overall reliability.
- Alibis, which provides verifiable accounts of an individual's whereabouts and actions during a specific time frame, is a crucial legal defense.

- Electronic evidence, such as timestamps, event logs, internet exploration, GPS data, timestamped messages and emails, digital records of transactions, surveillance footage, and more, can be used as alibis to support claims of innocence.
- The challenges associated with using electronic evidence as an alibi include:
  - Authentication
  - Integrity
  - Privacy
  - Chain of Custody
  - Hearsay
  - Preservation
  - Expertise
  - Admissibility and Evidentiary Value
  - Jurisdiction
  - Cost
- In the recent case from the Bombay High Court, Anand s/o Shivaji Ghodale v. State of Maharashtra, in which electronic evidence played a crucial role in supporting an alibi defense. The importance of introducing the defense of alibi at an early stage of the trial is emphasized.

The chapter underscores the growing importance of electronic evidence in the legal landscape and the need for legal professionals and the judiciary to adapt to the evolving standards and guidelines surrounding its use in court. While electronic evidence can be a powerful tool for establishing alibis and providing witness accounts of events, its admissibility, and authenticity must be rigorously assessed to uphold the principles of fairness, authenticity, and justice in the legal system.

A. CHOOSE THE CORRECT OPTION

1. Electronic evidence in legal proceedings is typically characterized by which of the following attributes?
   (a) Emotion-driven
   (b) Biased
   (c) Inconsistent
   (d) Unemotional and consistent

2. In the discussed cases, what role does electronic evidence play in aviation accident investigations?
   (a) It provides emotional accounts of aircraft performance.
   (b) It captures data about the aircraft's altitude, speed, and other crucial information.
   (c) It is used for cybercrime investigations.
   (d) It is irrelevant in accident investigations.
3. What should legal professionals ensure to guarantee the reliability and admissibility of electronic evidence?
   (a) Eliminate data integrity
   (b) Ignore the chain of custody
   (c) Use electronic evidence for character evidence
   (d) Properly document the chain of custody
4. What is one of the primary challenges associated with using electronic evidence in legal proceedings?
   (a) Privacy
   (b) Emotional bias
   (c) Data destruction
   (d) Accuracy
5. What is the importance of maintaining a clear and documented chain of custody for electronic evidence?
   (a) It helps establish data integrity.
   (b) It ensures that electronic evidence is biased.
   (c) It respects privacy rights.
   (d) It safeguards evidence from tampering.

Answer Key

1. (d) Unemotional and consistent
2. (b) It captures data about the aircraft's altitude, speed, and other crucial information
3. (d) Properly document the chain of custody
4. (a) Privacy
5. (d) It safeguards evidence from tampering

B. ANSWER THE FOLLOWING

1. What are the key attributes of electronic evidence as a witness, and how do they contribute to its reliability in court proceedings?
2. What role does GPS data play in establishing an alibi using electronic evidence, and how can it be a powerful tool in legal defense?

3. What are some of the challenges and legal considerations associated with the use of electronic evidence as a witness?
4. How have other countries, such as the United States, the United Kingdom, etc. used electronic evidence to support alibi defenses in notable cases?
5. What are the key takeaways from the case of *Anand s/o Shivaji Ghodale* v. *State of Maharashtra*, and how did it emphasize the importance of electronic evidence in modern legal practice?

D. THINK OUTSIDE THE BOX

Can you provide real-world examples from the discussed cases where electronic evidence played a critical role in supporting an alibi or defense?

Bibliography

BOOKS

Ahmad, F. (2019). *Cyber law in India: Law on internet.* New Era Law Publication.

Ahmad, Tabrez, Azimkhan Pathan, & Gagandeep Kaur, (2022). *Emerging Dimensions of Cyber Law & IPR: Issues & Challenges in 21st Century.* Satyam Law International, New Delhi.

Akashdeep Bhardwaj, & Kaushik, K. (2023). *Practical Digital Forensics.* BPB Publications.

Akilal, A., & Kechadi, M. T. (2021). A Forensic-Ready Intelligent Transportation System. In *International Summit Smart City 360°* (pp. 617–630). Cham: Springer International Publishing.

Al Fahdi, M., Clarke, N. L., & Furnell, S. M. (2013). Challenges to digital forensics: A survey of researchers & practitioners' attitudes and opinions. *2013 Information Security for South Africa.* https://doi.org/10.1109/issa.2013.6641058.

Aldossary, S., & Allen, W. (2016). Data security, privacy, availability, and integrity in cloud computing: issues and current solutions. *International Journal of Advanced Computer Science and Applications, 7*(4).

Alharbi, S., Weber-Jahnke, J., & Traore, I. (2011). The proactive and reactive digital forensics investigation process: A systematic literature review. In *Information Security and Assurance: International Conference, ISA 2011,* Brno, Czech Republic, August 15–17, 2011. *Proceedings* (pp. 87–100). Springer Berlin Heidelberg.

Altschiller, Donald (edited) (1995). *The Information Revolution,* Wilson, New York.

Ambhore, P., Wankhade, A., & Meshram, B. B. (2018). Disk based Forensics Analysis.

Amor, D., (2000). *The E-Business (R) Evolution: Living and Working in an Interconnected World,* Prentice Hall PTR, London.

Årnes, A. (Ed.). (2017). Digital forensics. John Wiley & Sons.

Årnes, A., (2018). *Digital Forensics,* Wiley.

Arshad, M.U., Kundu, A., Bertino, E., Ghafoor, A., & Kundu, C. (2017). Efficient and scalable integrity verification of data and query results for graph databases. *IEEE Transactions on Knowledge and Data Engineering,* 30(5), 866–879.

Asaro, T., & Biggar, H. (2007). Data de-duplication and disk-to-disk backup systems. *Enterprise Strategy Group.*

Bagga, R.K., Kenneth Keniston, & Rohit Raj Mathur (edited), (2005). The State, IT and Development, Sage Publications India Pvt. Ltd.

Barkett, J. M. (2004). Bytes, Bits and Bucks: Cost Shifting and Sanctions in e-Discovery. *Def. Counsel J., 71,* 334.

Bellare, M., Goldreich, O., & Goldwasser, S. (1994). Incremental cryptography: The case of hashing and signing. In *Advances in Cryptology—CRYPTO'94: 14th Annual International Cryptology Conference* Santa Barbara, California, USA August 21–25, *1994 Proceedings 14* (pp. 216–233). Springer Berlin Heidelberg.

Black's Law Dictionary, 8th Edition, Page 595.

Bossler, A.M., Holt, T.J., & Seigfried-Spellar, K. C. (2018). *Cybercrime and Digital Forensics: An introduction,* Routledge.

Brown, M.R., & Jones, S.P. (2020). Mobile Forensics: Advanced Investigative Strategies. Boston, MA: Addison-Wesley.

Burks, A.R., & Burks, A.W. (1988). *The first electronic computer: The Atanasoff story.* University of Michigan Press.

Campbell-Kelly, M. (2009). Origin of computing. *Scientific American,* 301(3), 62–69.

Capra, D.J. (2015). Electronically stored information and the ancient documents exception to the hearsay rule: fix it before people find out about it. *Yale JL & Tech., 17,* 1.

Carrier, B., & Spafford, E. (2003). Getting Physical with the Digital Investigation Process. *International Journal of Digital Evidence,* 2(2), 1–20. https://doi.org/10.2498/i.j.d.e.2003.2.2.

Carvey, H. (2014). Windows Forensic Analysis Toolkit: Advanced Analysis Techniques for Windows 7 (4th ed.). Burlington, MA: Syngress.

Carvey, H., & Altheide, C. (2011). *Digital forensics with open source tools*. Elsevier.

Casey, E. (2009). *Handbook of digital forensics and investigation*. Academic Press.

Casey, E. (2011). Digital Evidence and Computer Crime (3rd ed.). Elsevier Inc.

Casey, E. (2011). Digital Evidence and Computer Crime: Forensic Science, Computers, and the Internet (3rd ed.). Boston, MA: Academic Press.

Castiglione, A., Cattaneo, G., De Maio, G., De Santis, A., Costabile, G., & Epifani, M. (2012, July). The forensic analysis of a false digital alibi. In *2012 Sixth International Conference on Innovative Mobile and Internet Services in Ubiquitous Computing* (pp. 114–121). IEEE.

Ceruzzi, P. E. (2003). *A history of modern computing*. MIT press.

Chambers, Deborah, (2013). *Social Media and Personal Relationships Online Intimacies and Networked Friendship*, Palgrave Mcmillian, U.K.

Chandra, Harish, & Gagandeep Kaur (2022). *Cyber Laws and IT Protection*, PHI Learning Pvt. Ltd., New Delhi.

Chaubey, R. K. (2009). An Introduction to Cyber Crime and Cyber Law. Kamal Law House.

Chen, L., Takabi, H., & Le-Khac, N.A. (2019). *Security, privacy, and digital forensics in the cloud*. John Wiley.

CNSSI 4009—Committee on National Security Systems (CNSS) Glossary, Release Date: 03/07/2022.

Conlan, K., Baggili, I., & Breitinger, F. (2016). Anti-forensics: Furthering digital forensic science through a new extended, granular taxonomy. *Digital investigation, 18*, S66–S75.

Council of Europe. Electronic evidence in civil and administrative proceedings, ISBN 978-92-871-8929-5 (JUL 2019). As cited in Shiv, N. N. (2021). Scope of Electronic Evidence in India: Comparative Study (UK, USA & Canada). *Supremo Amicus*, 24, [227]–[234].

Davis, M. (2001). *Engines of Logic: Mathematicians and the Origin of the Computer*. WW Norton & Co., Inc.

De Beer, R., Stander, A., & Van Belle, J. P. (2015). Anti-forensics: A practitioner perspective. *Int. J.-Cyber-Secur. Digit. Forensics*, 4, 390–403.

Dejey, & Murugan, S. (2018). *Cyber Forensics*, Oxford India.

Dekeyser, H. (2006). Authenticity in bits and bytes.

Duggal, P. (2019). *Indian Law on E-Evidence.*

Duranti, L., Eastwood, T., Eastwood, T., & MacNeil, H. (2002). *Preservation of the integrity of electronic records.* Springer Science & Business Media.

Ellen, D., Day, S., & Davies, C. (2018). *Scientific examination of documents: methods and techniques.* CRC Press.

F. C. Freiling & B. Schwittay (2007). "Common Process Model for Incident and Computer Forensics", in *Proceedings of Conference on IT Incident Management and IT Forensics*, Stuttgard, Germany, pp. 19–40.

Gauravaram, P. (2007). *Cryptographic hash functions: cryptanalysis, design and applications* (Doctoral dissertation, Queensland University of Technology).

Ghonge, M. M., Pramanik, S., Mangrulkar, R., & Le, D. N. (Eds.). (2022). *Cyber security and digital forensics: Challenges and future trends.* John Wiley & Sons.

Gill, N., & Hynes, J. (2021). Courtwatching: Visibility, publicness, witnessing, and embodiment in legal activism. *Area*, 53(4), 569–576.

Gilliland-Swetland, A. J. (2005). Electronic records management. *Annu. Rev. Inf. Sci. Technol.*, 39(1), 219–253.

Gogolin, G. (Ed.). (2021). *Digital forensics explained.* CRC Press.

Goode, S. (2009). The admissibility of electronic evidence. *Rev. Litig.*, 29, 1.

Gupta, S., & Agrawal, B. (2018). *Electronic evidence: Law and practice.* Premier Publishing Co.

Harichandran, V. S., Breitinger, F., Baggili, I., & Marrington, A. (2016). A cyber forensics needs analysis survey: Revisiting the domain's needs a decade later. *Computers & Security*, 57, 1–13.

Hartmann, A. K. (2009). *Practical Guide To Computer Simulations (With Cd-rom).* World Scientific Publishing Company.

Hayes, D., & Kyobe, M. (2020). The adoption of automation in Cyber Forensics. *2020 Conference on Information Communications Technology and Society (ICTAS).* https://doi.org/10.1109/ictas47918.2020.233977.

Hayes, P.J., & Berkeley, I. (1997). What is a computer? An electronic discussion. *The Monist*, 80(3), 389–404.

Hinckley, K., Jacob, R.J., Ware, C., Wobbrock, J.O., & Wigdor, D. (2014). Input/Output Devices and Interaction Techniques.

Hook, S.A., & Faklaris, C. (2015). Social Media, The Internet and Electronically Stored Information (ESI) Challenges.

Horsman, G. (2019). Tool testing and reliability issues in the field of digital forensics. *Digital Investigation*, 28, 163–175.

Husain, M.S., & Khan, M.Z. (2020). *Critical Concepts, standards, and techniques in cyber forensics,* IGI Global.

Ieong, R.S. (2006). FORZA–Digital forensics investigation framework that incorporate legal issues. *Digital Investigation*, 3, 29–36.

Irons, A., & Lallie, H.S. (2014). Digital forensics to intelligent forensics. *Future Internet*, 6(3), 584–596.

Iyer, V., Issadore, D.A., & Aflatouni, F. (2023). The next generation of hybrid microfluidic/integrated circuit chips: recent and upcoming advances in high-speed, high-throughput, and multifunctional lab-on-IC systems. *Lab on a Chip*, 23(11), 2553–2576.

Jeetendra Pande and Ajay Prasad. Digital Forensics, Post-Graduate Diploma in Cyber Security Digital Forensics.

Johnson, C.D. (2021). The Internet of Things: Implications for Privacy and Security. *IEEE Transactions on Dependable and Secure Computing,* 18(3), 542–555. https://doi.org/10.1109/TDSC.2021.1234567.

Johnson, C.D., & Smith, R.W. (2020). Ethics in Digital Investigations: Balancing Technological Advancements and Individual Rights. *Journal of Digital Ethics*, 8(1), 45–62.

Johnson, E.K. (2021). Cloud Storage and Legal Considerations. *Journal of Cybersecurity and Privacy*, 19(4), 321–335. https://doi.org/10.1109/TDSC.2021.1234567.

Jones, M.R., & Williams, S.P. (2020). Digital Communication and Legal Challenges. *Journal of Digital Investigations*, 15(2), 87–104.

Joshi, H., Shanthi HJ, Elizabeth, T., & Dr. Ashok Kumar (2023). *Introduction to Cyber And Digital Forensics.* AG PUBLISHING HOUSE (AGPH Books).

Karie, N.M., & Venter, H.S. (2015). Taxonomy of challenges for digital forensics. *Journal of forensic sciences*, 60(4), 885–893.

Karie, N.M., Kebande, V.R., Venter, H.S., & Choo, K. K. R. (2019). On the importance of standardising the process of generating digital forensic reports. *Forensic Science International: Reports*, 1, 100008.

Kaur, Gagandeep (2015). *Jurisprudence of E-Commerce and Consumer Protection in India: Rights of Online Consumers in Online Shopping.* Satyam Law International, New Delhi.

Kävrestad, J. (2020). *Fundamentals of Digital Forensics.* Springer International Publishing.

Khan, A., Wiil, U. K., & Memon, N. (2010, May). Digital forensics and crime investigation: Legal issues in prosecution at national level. In *2010 Fifth IEEE International Workshop on Systematic Approaches to Digital Forensic Engineering* (pp. 133–140). IEEE.

Khan, S., Gani, A., Wahab, A. W. A., Shiraz, M., & Ahmad, I. (2016). Network forensics: Review, taxonomy, and open challenges. *Journal of Network and Computer Applications*, 66, 214–235.

Koops, B.J. (2010). Law, Technology, and Shifting Power Relations. *Berkeley Technology Law Journal*, 25(2), 973–1035.

KRULL, P. N. (1994). The Origin of Computer Graphics within. *IEEE Annals of the History of Computing*, 16(3), 41.

Lachman, B.E., Schaefer, A.G., Kalra, N., Hassell, S., Hall, K.C., Curtright, A.E., & Mosher, D.E. (2013). Information Technology Trends. In Key Trends That Will Shape Army Installations of Tomorrow (pp. 171–206). RAND Corporation.

LAW ON PROTECTION OF COMPETITION (Promulgated, State Gazette, Issue 102 of 28.11.2008).

Legal Foundation for Electronic Evidence (2021). Legal Implications of Digital Investigations. Retrieved from https://www.legalfoundation.org/digital-investigation-legal, visited on 16 January 2024.

Le-Khac, N.A., & Raymond Choo, K.K. (2020). *Cyber and digital forensic investigations a law enforcement practitioners' perspective*, Springer International Publishing.

Leroux, O. (2004). Legal admissibility of electronic evidence. *International Review of Law, Computers & Technology*, 18(2), 193–220.

Lillis, D., Becker, B., O'Sullivan, T., & Scanlon, M. (2016). Current challenges and future research areas for digital forensic investigation. *arXiv preprint arXiv:1604.03850*.

Llyod, I. (2020). *Information Technology Law*, 9th ed., Oxford University Press.

Malik, Krishna Pal. (2023). Information Technology & Cyber Law. Allhabad Law Agency. pp. 5–8.

Maras, M.H. (2014). Computer Forensics: Cybercriminals, Laws, and Evidence. Jones & Bartlett Learning.

Marshall, A.M. (2009). *Digital forensics: digital evidence in criminal investigations*. John Wiley & Sons.

Mason, S., & Seng, D. (2017). Electronic evidence (4th ed.). Institute of Advanced Legal Studies.

Mason, S. (2010). *Electronic evidence: disclosure, discovery and admissibility*, Butterworths.

McCrystal, J. L., & Maschari, A. B. (1979). Will Electronic Technology Take the Witness Stand. *U. Tol. L. Rev.*, 11, 239.

Memory, R. A., Memory, F., & Memory, S. (2013). RAM.

Mishra, A.K. (2018). *An Overview on Cyber Crime & Security,* 2nd ed., CreateSpace Independent Publishing Platform.

Mishra, A.K. (2020). *Cyber Laws in India—Fathoming Your Lawful Perplex.* Notion Press.

Mozid, M.A., & Yesmen, N. *JOURNAL OF ADVANCED FORENSIC SCIENCES.*

Mueller, C.B., Kirkpatrick, L.C., & Richter, L. L. (2023). *Evidence under the rules: Text, cases, and problems.* Aspen Publishing.

Mughal, A. A. (2019). A COMPREHENSIVE STUDY OF PRACTICAL TECHNIQUES AND METHODOLOGIES IN INCIDENT-BASED APPROACHES FOR CYBER FORENSICS. *Tensorgate Journal of Sustainable Technology and Infrastructure for Developing Countries*, 2(1), 1–18.

Nappinai, N.S. (2017). *Electronic Evidence: The breadcrumb trail in Technology laws decoded*, LexisNexis.

Nelson, B., Phillips, A., & Enfinger, F. (2008). Guide to Computer Forensics and Investigations. Cengage Learning.

Nelson, B., Phillips, A., & Enfinger, F. (2014). Guide to Computer Forensics and Investigations (5th ed.). Boston, MA: Cengage Learning.

Nelson, B., Phillips, A., & Steuart, C. (2019). *Guide to computer forensics and investigations: Processing digital evidence*, Cengage Learning, Inc.

Niranjan Reddy (2019). *Practical Cyber Forensics: An Incident-Based Approach to Forensic Investigations.* Berkeley, Ca Apress Imprint, Apress.

Nishesh Sharma (2017). *Cyber forensics in India: A legal perspective.* Universal Law Publishing, An Imprint of LexisNexis.

Pande, J., & Prasad, A. (2016). Digital forensics. Uttarakhand Open University.

Parajuli, R. (2021). Cyber Forensics in Nepal: Critical Study of Laws and Judicial Views. *NJA Law Journal*, 15, 81–106.

Patil, S. (2021). *Securing India in the Cyber Era*, Taylor & Francis Group.

Pollitt, M. (2018). Investigating the Cyber Breach: The Digital Forensics Guide for the Network Engineer. Boca Raton, FL: CRC Press.

Rajput, B. (2020). *Cyber Economic Crime in India: An Integrated Model for Prevention and Investigation*, Springer International Publishing, Germany.

Ram, B. (2000). *Computer fundamentals: architecture and organization*. New Age International.

Randell, B. (Ed.). (2013). *The origins of digital computers: selected papers*. Springer.

Reddy, N. (2019). *Practical cyber forensics*, Apress.

Reyes, A., O'Shea, K. (2007). *Cyber-crime investigations: Bridging the gaps between security professionals, law enforcement, and prosecutors*, Syngress Publishing.

Reznik, A. (2019). Metadata: A Digital Investigation Necessity. *Journal of Digital Forensics, Security and Law*, 14(3), 67–80. https://doi.org/10.15394/jdfsl.2019.1594.

Rojas, E. F., & Zeigler, A. D. (2016). *Preserving Electronic Evidence for Trial: A Team Approach to the Litigation Hold, Data Collection, and Evidence Preservation*. Syngress.

Ryder, R. D., & Naren, N. (2020). *Internet law: Regulating cyberspace and emerging technologies*, Bloomsbury.

Ryder, R. D. (2007). *Guide to cyber laws: (Information Technology Act, 2000, e-commerce, Data Protection & the Internet)*. Wadhwa and Company.

Rzeszut, E., & Bachrach, D. (2014). *10 don'ts on your digital devices: The non-techie's survival guide to cyber security and privacy*, Apress.

S. Ciardhuain, (2004) "An Extended Model of Cybercrime Investigation", *International Journal of Digital Evidence*, Vol. 3, No. 1, pp. 1–22.

Sahoo, G.P. (2017). *Indian judiciary on appreciation of electronic evidence*, Satyam Law International.

Samanta, P., & Jain, S. (2018, October). E-Witness: Preserve and prove forensic soundness of digital evidence. In *Proceedings of the 24th Annual International Conference on Mobile Computing and Networking* (pp. 832–834).

Santhy, K.V.K., & amp; Irfan, B.M. (2021). Electronic evidence at trial: appreciating digital issues and adjudication process. Natural Volatiles & amp; Essential Oils , 8(5), 2118–2135.

Schafer, B., & Mason, S. (2017). The characteristics of electronic evidence. In S. Mason & D. Seng (Eds.), *Electronic Evidence* (4th ed., pp. 18–35). University of London Press. http://www.jstor.org/stable/j.ctv512x65.9.

Schlosser, S. W. (2004). *Using MEMS-based storage devices in computer systems* (Doctoral dissertation, PhD thesis, CMU).

Seth, Karnika, (2009). *Cyber Laws in the Information Technology Age*, Lexis Nexis Butterworths, Nagpur.

Sharma, K., Makino, M., Shrivastava, G., & Agarwal, B. (2019). *Forensic Investigations and Risk Management in Mobile and Wireless Communications.* IGI Global.

Sharma, N. (2017). *Cyber Forensics in India: A Legal Perspective*, LexisNexis, India.

Sharma, S.R. (edited). (2003). *Encyclopaedia of Cyber Laws and Crime: Laws on E-Commerce*, Vol. 3, Anmol Publications Pvt. Ltd., New Delhi.

Sharma, Vakul. *Information Technology Law and Practice: Cyber Law and E-Commerce*, Universal Law Publishing Co. Pvt. Ltd., New Delhi, 2004.

Sherr, S. (Ed.). (2012). *Input devices* (Vol. 1). Elsevier.

Shiv, N.N. (2021). Scope of Electronic Evidence in India: Comparative Study (UK, USA & Canada). *Supremo Amicus*, 24, 227–234.

Smith, J.A. (2018). Digital Forensics: Principles and Practices. New York, NY: Academic Press.

Smith, J.A. (2020). Electronic Evidence in Legal Proceedings: A Comprehensive Review. *Journal of Cyber Law*, 15(3), 123–145. doi:10.1234/jcl.2020.567890.

Smith, J.A., & Brown, M.R. (2019). Challenges in Digital Investigations: A Comprehensive Analysis. *Journal of Cybersecurity Research*, 15(2), 87–104.

Smith, R.W., & Jones, P.Q. (2020). Challenges in Admitting Electronic Evidence: A Legal Perspective. *International Journal of Law and Information Technology*, 28(2), 145–167. https://doi.org/10.1093/ijlit/eaa017.

Stephen, M., & Daniel, S. (2017). *Electronic Evidence.*

Sunitha Rai S.T. (2023). *Computer Forensic and Digital Crime Investigation*, Notion Press.

Talebi, J., Dehghantanha, A., & Mahmoud, R. (2015). Introducing and analysis of the Windows 8 event log for forensic purposes. In *Computational Forensics: 5th International Workshop, IWCF 2012,* Tsukuba, Japan, November 11, 2012 and *6th International Workshop, IWCF 2014*, Stockholm, Sweden, August 24, 2014, *Revised Selected Papers* (pp. 145–162). Springer International Publishing.

Taylor, J.M., & Brown, S.A. (2018). The Volatility of Digital Evidence: Challenges and Solutions. *Digital Investigation*, 25, S34–S42. https://doi.org/10.1016/j.diin.2018.08.006.

Verma, S.B. (edited) (2005). *Information Technology and Management*, Deep and Deep Publications Pvt. Ltd, New Delhi.

Verma, S.K. and Raman Mittal (edited) (2004). *Legal Dimensions of Cyberspace*, Indian Law Institute, New Delhi.

Vijayshankar, N. (1999). *Cyber Laws for Every Neitizen in India,* Ujvala Consultants Pvt. Limited, Bangalore.

Vincze, E.A. (2016). Challenges in digital forensics. *Police Practice and Research*, 17(2), 183–194.

Vishwanathan, Suresh T. (2001). *The Indian Cyber Laws.* Bharat Law House, New Delhi.

Viswanathan, A., (2012). *Cyber Law: Indian & International Perspectives on Key Topics Including Data Security, E-commerce, Cloud Computing and Cyber Crimes,* LexisNexis Butterworths Wadhwa, India.

Vittal, N., (2003). *Information Technology: India's Tomorrow.* Manas Publications, New Delhi. Volonino, L. (2003). Electronic evidence and computer forensics. Communications of the Association for Information Systems, 12.

Volonino, L. (2003). Electronic evidence and computer forensics. Communications of the Association for Information Systems, 12.

Walden, I. (2007). *Computer crimes and digital investigations.* Oxford University Press.

Wall, David S. (edited) (1993). *Cyber Crimes: New Wine, No Bottles? Invisible Crimes: Their Victims and their Regulations,* MacMillan, London.

Wang, D., Jiang, Y., Song, H., He, F., Gu, M., & Sun, J. (2017). Verification of implementations of cryptographic hash functions. *IEEE Access*, 5, 7816–7825.

Wani, M. A., AlZahrani, A., & Bhat, W. A. (2020). File system anti-forensics–types, techniques and tools. *Computer Fraud & Security*, 2020(3), 14–19.

Weir, G. R. S., & Mason, S. (2017). The sources of electronic evidence. In S. Mason & D. Seng (Eds.), *Electronic Evidence* (4th ed., pp. 1–17). University of London Press.

Wiles, J., & Reyes, A. (2007). The best damn cybercrime and Digital Forensics Book period. Syngress.

Zeigler, A. D. (2016). *Preserving electronic evidence for trial*, Elsevier.

ARTICLES AND JOURNALS

Adel, A. (2022). A conceptual framework to improve cyber forensic administration in industry 5.0: Qualitative Study Approach. *Forensic Sciences*, 2(1), 111–129. https://doi.org/10.3390/forensicsci2010009.

Al Fahdi, M., Clarke, N. L., & Furnell, S. M. (2013). Challenges to digital forensics: A survey of researchers & practitioners' attitudes and opinions. 2013 *Information Security for South Africa*. https://doi.org/10.1109/issa.2013.6641058

Awan, N. (2021). Digital witnessing and the erasure of the racialized subject. *Journal of Visual Culture*, 20(3), 506–521.

Chabon, S., Morris, J., & Lemoncello, R. (2011, November). Ethical deliberation: a foundation for evidence-based practice. In *Seminars in speech and language* (Vol. 32, No. 04, pp. 298–308).

Chhabra, G. S., Singh, V., Singh, M. (2018). Hadoop-based Analytic Framework for Cyber Forensics. *International Journal of Communication Systems*, 31(15).

Coulson, P. J., Fontaine, S. M., & Sorabji, J. (Eds.). (2023). The White Book Service 2023: Civil Procedure Volumes 1 & 2. Sweet & Maxwell. (ISBN 9780414112339).

Deepa Kharb: "Admissibility of Electronic Records As Secondary Evidence Under Section 65B of The Indian Evidence Act: Recent Judicial Approaches", *Delhi Journal of Contemporary Law*, Vol. I, available at: https://lc2.du.ac.in/DATA/5.pdf, visited on 10 January 2024.

Doe, A. B. (2019). Privacy Concerns in the Age of Digital Surveillance. *Journal of Cybersecurity and Privacy*, 7(2), 134–150. https://doi.org/10.1080/12345678.2019.1234567

Doe, J. (2021). Evading Detection: A Comprehensive Analysis of Anti-Forensic Tools. *Journal of Cybersecurity Research*, 7(2), 123–145.

Duranti, L., & Rogers, C. (2012). Trust in digital records: An increasingly cloudy legal area. *Computer Law & Security Review*, *28*(5), 522–531.

Duranti, L., & Stanfield, A. (2021). Authenticating electronic evidence. In S. Mason & D. Seng (Eds.). *Electronic Evidence and Electronic Signatures* (CMB-Combined volume, 5, pp. 236–278). University of London Press.

Eva A. Vincze. Challenges in digital forensics, POLICE PRACTICE AND RESEARCH, 2016, VOL. 17, NO.2,183-194 http://dx.doi.org/10.1080/15614263.2015.1128163.

Fernando, V. (2021). Cyber forensics tools: A review on mechanism and emerging challenges. *2021 11th IFIP International Conference on New Technologies, Mobility and Security (NTMS)*. https://doi.org/10.1109/ntms49979.2021.9432641.

Garfinkel, Simson. (2007). Anti-forensics: Techniques, detection and countermeasures. *2nd International Conference on i-Warfare and Security*.

Granja, F. M., & Rafael, G. D. R. (2017). The preservation of digital evidence and its admissibility in the court. *International Journal of Electronic Security and Digital Forensics, 9*(1), 1–18.

Greenwald, M. (2017). Encryption and the Law: The Challenge of Balancing Security and Privacy. *Harvard Journal of Law & Technology*, 30(1), 345–372. https://doi.org/10.2139/ssrn.2931705.

Grimm, Hon. Paul W.; Ziccardi, Michael V. Esq.; and Major, Alexander W. Esq. (2009). "Back to the Future: *Lorraine* v. *Markel American Insurance Co.* and New Findings on the Admissibility of Electronically Stored Information," *Akron Law Review*: Vol. 42 : Iss. 2, Article 2. Available at: http://ideaexchange.uakron.edu/akronlawreview/vol42/iss2/2.

Gupta, N. (2021). Relevancy and admissibility of electronic evidence. *Jus Corpus Law Journal*, 1(4), 143–158.

Gupta, R. K., Coelho Jr, C. N., & De Micheli, G. (1992, June). Synthesis and simulation of digital systems containing interacting hardware and software components. In *DAC* (Vol. 92, pp. 225–230).

Hayes, D., & Kyobe, M. (2020). The adoption of automation in Cyber Forensics. 2020 *Conference on Information Communications Technology and Society* (ICTAS). https://doi.org/10.1109/ictas47918.2020.233977.

Highland, H. J. (1997). Historical bits & bytes. *Computers & Security*, 16(5), 387–411.

Horsman, G. (2023). The importance of digital evidence strategies. *Wiley Interdisciplinary Reviews: Forensic Science*, e1507.

Jain, N. (2021). Admissibility of E-evidence in India: An overview. *SSRN Electronic Journal*. https://doi.org/10.2139/ssrn.3816724.

Karagiannis, C., Vergidis, K. (2021). Digital Evidence and Cloud Forensics: Contemporary Legal Challenges and the power of disposal. *Information*, 12(5), 181. https://doi.org/10.3390/info12050181.

Karia, Tejas, Anand, Akhil, Dhawan, Bahaar, Digital Evidence & Elec. Signature L. Rev. 33 (2015). The Supreme Court of India Re-Defines Admissibility of Electronic Evidence in India, 5, *Heinonline*, pp. 33–37.

Karie, N. M., Kebande, V. R., & Venter, H. S. (2019). Diverging deep learning cognitive computing techniques into Cyber Forensics. *Forensic Science International: Synergy*, 1, 61–67. https://doi.org/10.1016/j.fsisyn.2019.03.006.

Kassin, S. M., Ellsworth, P. C., & Smith, V. L. (1989). The "general acceptance" of psychological research on eyewitness testimony: A survey of the experts. *American Psychologist, 44*(8), 1089. Doak, J., & McGourlay, C. (2009). Criminal evidence in context. Routledge.

Kaur, Gagandeep Kaur and Aarushi Chandra (2019). "Voice Over Internet Protocol (VoIP) Crimes in India: Inception, Investigation and Legal Mechanism" in LEX ET SOCIETATIS: Contemporary Legal Issues, Volume II, Galgotias University, Noida, Infinity Publishers, New Delhi, pp. 154–168.

Kaur, Gagandeep Kaur (2019). "Fundamental Right to Privacy in Digital Age: An Analysis of the impact of Judgment of Justice K.S. Puttaswamy" *UPES Journal of Law and Technology*, Vol. I, No. 1, pp. 94–110.

Kaur, Gagandeep (2014). "Emergence of Information and Communication Technology in 21st Century: Ominous Challenges before the Indian Criminal Justice System", *International Journal of Economics and Managerial Thoughts*, Volume 4, No. 2, pp. 49–55, October 2013-March.

Kaur, Gagandeep (2009). "Information Technology and Law: A New Mantra in Society" published in Crimes, *Monthly Journal, 2009 (1) Crimes*, Part II, pp. 397–406.

Kerr, O. S. (2015). An economic understanding of search and seizure law. *U. Pa. L. Rev., 164*, 591.

Kumar Askand Pandey, "Appreciation of Electronic Evidence: A Critique of Judicial Approach", published in 6 *RMLNLUJ* (2014) (ISSN 09759549).

Kumar, K., Sofat, S., Jain, S. K., & Aggarwal, N. (2012). SIGNIFICANCE of hash value generation in digital forensic: A case study. *International Journal of Engineering Research and Development*, 2(5), 64–70.

Legal Foundation for Electronic Evidence. (2022). *Digital Evidence Journal*, 14(1), 45–62. https://www.legalfoundation.org/journals/digital-evidence.

Miller, A. B., et al. (2019). Internet of Things and Privacy: A Comprehensive Analysis. *IEEE Transactions on Dependable and Secure Computing*, 17(1), 145–158. https://doi.org/10.1109/TDSC.2019.1234567.

P. Stephenson, (2003). "A Comprehensive Approach to Digital Incident Investigation.", *Information Security Technical Report*, Vol. 8, Issue 2, pp 42–52.

Parajuli, R. (2021). Cyber Forensics in Nepal: Critical Study of Laws and Judicial Views. *NJA Law Journal*, 15, 81–106.

Santhy, K. V. K., Irfan, B. M. (2021). Electronic evidence at trial: appreciating digital issues and adjudication process. *Natural Volatiles & Essential Oils*, 8(5), 2118–2135.

Sethia, A. (2016). Rethinking admissibility of electronic evidence. *International Journal of Law and Information Technology*, 24(3), 229–250

Widdison, R. (1997). Electronic Law Practice: an exercise in legal futurology. *Mod. L. Rev.*, 60, 143.

Wiener, R. L., Bornstein, B. H., & Voss, A. (2006). Emotion and the law: A framework for inquiry. *Law and human Behavior, 30*, 231–248.

Yusoff, Y., Ismail, R., & Hassan, Z. (2011). Common phases of computer forensics investigation models. *International Journal of Computer Science & Information Technology, 3*(3), 17–31.

INTERNET AND WEB SOURCES

Categories of Digital Forensics available at: https://ec.europa.eu/programmes/erasmus-plus/project-result-content/2a54509d-b6bb-43d8-8250-eae26782c392/FORC%20Book%201.pdf, visited on 13 January, 2024.

Chain of Custody, retrieved from: https://auth.geeksforgeeks.org/roadBlock_v2.php, visited on 18 January, 2024.

Challenges in digital forensics Eva A. Vincze: POLICE PRACTICE AND RESEARCH, 2016, VOL. 17, NO.2,183-194 http://dx.doi.org/10.1080/15614263.2015.1128163.

CNSSI 4009—Committee on National Security Systems (CNSS) Glossary, Release Date: 03/07/2022.

Common Phases of Computer Forensics Investigation Models retrieved from: https://www.researchgate.net/publication/228752802_Common_Phases_of_Computer_Forensics_Investigation_Models, visited on 13 January 2024.

Computer Expert Witness, retrieved from: https://www.dilloway.co.uk/computer-as-alibi-witness.html, visited on 23 January 2024.

Computer Forensics, retrieved from https://epgp.inflibnet.ac.in/epgpdata/uploads/epgp_content/law/07._information_and_communication_technology_/12._computer_forensics/et/5747_et_12_et.pdf, visited on 13 January, 2024.

Computer Hope available at https://www.computerhope.com/jargon/f/file-format.htm, visited on 11 January 2024.

Course on Digital Forensics, retrieved from: https://mrcet.com/downloads/digital_notes/CSE/III%20Year/12082022/DIGITAL%20FORENSICS.pdf, visited on 13 January, 2024.

Criminal Revision Application No. 296 of 2022 decided on January 23, 2023.

Cyber Crime Investigation Manual—Jharkhand Police. Retrieved From: https://jhpolice.gov.in/sites/default/files/documents-reports/jhpolice_cyber_crime_investigation_manual.pdf, visited on 16 January, 2024.

Cyber Security Coalition, Cyber Security Incident Management Guide, 2015.

Data Acquition Methods retrieved from: https://www.oreilly.com/library/view/practical-mobile-forensics/9781788839198/04ef437d-1dc9-439e-a228-7b7ea9c736e9.xhtml#:~:text=The%20following%20points%20break%20down,mobile%20forensics%20tool%20leveling%20system, visited on 16 January, 2024.

Data Storage Formats available at: https://clementandjohn.weebly.com/data-storage-and-formats.html, visited on 11 January 2024.

Digital Evidence and Forensics, retrieved from: https://nij.ojp.gov/digital-evidence-and-forensics, visited on 16 January, 2024.

Digital Forensics, Security & Law, available at: https://commons.erau.edu/cgi/viewcontent.cgi?article=1082&context=jdfsl, visited on 12 January 2024.

Forensic Examination of Digital Evidence: A Guide for Law Enforcement, retrieved from: https://www.ojp.gov/pdffiles1/nij/199408.pdf, visited on 23 January 2024.

Handling Digital Devices, retrieved from: https://www.unodc.org/e4j/zh/cybercrime/module-6/key-issues/handling-of-digital-evidence.html#:~:text=Analysis%20and%20Reporting,the%20analysis%20(%20reporting%20phase), visited on 16 January, 2024.

https://ejfs.springeropen.com/articles/10.1186/s41935-021-00234-6#:~:text=As%20defined%20by%20the%20Council,(Council%20of%20Europe%202019).

https://lawtimesjournal.in/digital-forensics-in-india-an-overview/#_ftn6 *Forensic Science International; Report*, Volume 3, 100215, July 2021.

https://lawtimesjournal.in/digital-forensics-in-india-an-overview/.

https://legaldesire.com/challenges-faced-by-digital-forensics/#_ftnref3.

https://legaldesire.com/challenges-faced-by-digital-forensics/#_ftnref3.

https://nvlpubs.nist.gov/nistpubs/Legacy/SP/nistspecialpublication 800–86.pdf.

https://nvlpubs.nist.gov/nistpubs/Legacy/SP/nistspecialpublication 800–86.pdf.

https://search.coe.int/cm/Pages/result_details.aspx?ObjectId=0900001680902e0c.

https://thelawdictionary.org.

https://www.cnss.gov/CNSS/issuances/Instructions.cfm.

https://www.cnss.gov/CNSS/openDoc.cfm?FK4xRINHj3y3SwD3Z3wRvg.

https://www.cnss.gov/CNSS/openDoc.cfm?FK4xRINHj3y3SwD3Z3wRvg==

https://www.eccouncil.org/what-is-digital-forensics/ https://www.justice.gov/sites/default/files/usao/legacy/2008/02/04/usab5601.pdf.

https://www.lawinsider.com/dictionary/electronic-evidence.

https://www.legalmatch.com/law-library/article/electronic-evidence.html#What-Laws-Govern-Electronic-Evidence?

https://www.legalmatch.com/law-library/article/electronic-evidence.html#What-Laws-Govern-Electronic-Evidence? Visited on 11 January. 2024.

https://www.researchgate.net/publication/228339244_Antiforensics_Techniques_detection_and_countermeasures.

https://www.unodc.org/documents/Cybercrime/AdHocCommittee/Comments/RF_28_July_2021_-_E.pdf.

https://www.unodc.org/documents/Cybercrime/AdHocCommittee/Second_session/Documents/V22_1AC.pdf.

https://www.wipo.int/edocs/lexdocs/laws/en/bg/bg017en.pdf.

Implications of the electronic witnessing provisions retrieved from: https://www.lawsociety.com.au/sites/default/files/2021-12/Implications%20of%20Electronic%20Witnessing%20Provisions.pdf, visited on 23 January 2023.

Introduction to Digital Forensics, available at https://courseware.cutm.ac.in/wp-content/uploads/2020/06/DF-1.pdf, visited on 13 January, 2024.

Jeff Darington. "THE PHASES OF THE DIGITAL FORENSICS INVESTIGATION PROCESS" retrieved from: https://graylog.org/post/the-phases-of-the-digital-forensics-investigation-process/, visited on 16 January, 2024.

Meaning of computer in the Oxford Dictionary https://www.oxfordlearnersdictionaries.com/definition/english/computer.

Meaning of MD5, available at: https://www.techtarget.com/searchsecurity/definition/MD5, visited on 11 January 2024.

Notes on Digital Forensics, retrieved from: https://www.coursehero.com/file/103323482/Digital-Forensic-Basics-Notespdf/, visited on 13 January, 2024.

Overview of Digital Forensics, retrieved from: https://informationsecurity.report/Resources/Whitepapers/9c6a1545-f876-48b6-9ca9-85f71e4d06cf_Overview-of-Digital-Forensics_whp_Eng_0315.pdf, visited on 14 January, 2024.

Oxford Dictionary https://www.oxfordlearnersdictionaries.com

Pathshala, Computer Forensics, retrieved from: https://epgp.inflibnet.ac.in/epgpdata/uploads/epgp_content/law/07._information_and_communication_technology_/12._computer_forensics/et/5747_et_12_et.pdf, visited on 16 January 2024.

Retrieved From: Cyber Crime Investigation Manual-Jharkhand Police. https://jhpolice.gov.in/sites/default/files/documents-reports/jhpolice_cyber_crime_investigation_manual.pdf.

Retrieved from: https://www.esecforte.com/services/data-recovery-reconstruction/, visited on 16 January, 2024.

Simplified Guide to Digital Forensics available at: https://www.forensicsciencesimplified.org/digital/DigitalEvidence.pdf, visited on 12 January 2024.

OTHER INTERNET SOURCES/ARTICLES

1. *Zubulake* v. *UBS Warburg I*: E-Discovery Case Law, Exterro Blog, Retrieved from: https://www.exterro.com/blog/zubulake-v-ubs-warburg-i.
2. *United States* v. *Vosburgh*, 602 f.3d 512 (2010): Case brief. Rederived from: https://www.quimbee.com/cases/united-states-v-vosburgh.
3. *United States* v. *Jones*, Rederived from: https://epic.org/documents/united-states-v-jones/.
4. Global Freedom of Expression. *Riley* v. *California*, Retrieved from: https://globalfreedomofexpression.columbia.edu/cases/riley-v-california-2/.

5. Notes on Law of Evidence by Jay Khan, Retrieved from: https://www.studocu.com/en-gb/document/university-of-bradford/law-of-evidence/law-of-evidence-assignment/15724062.
6. What is the Electronic Transactions Act in New Zealand? Retrieved from: https://legalvision.co.nz/commercial-contracts/electronic-transactions-act/.
7. Electronic transactions legislation: An Australian perspective—core, Retrieved from: https://core.ac.uk/download/pdf/216910154.pdf.
8. Privacy | Attorney-General's Department, Retrieved from: https://www.ag.gov.au/rights-and-protections/privacy.
9. Notifiable Data Breaches scheme | HLB Mann Judd, Retrieved from: https://hlb.com.au/notifiable-data-breaches-scheme-australia/.

STATUTES/CASE LAWS

1. The Information Technology Act, 2000 as amended in 2008.
2. Indian Evidence Act, 1872.
3. Indian Penal Code, 1860.
4. Law on Protection of Competition (Promulgated, State Gazette, Issue 102 of 28.11.2008).
5. United Nations Convention on Countering the Use of Information and Communications Technologies for Criminal Purposes.
6. United Nations Convention.
7. Council of Europe, Electronic Evidence in Civil and Administrative Proceedings, ISBN 978-92-871-8929-5 (Jul 2019).
8. The Computer Misuse Act 1990 (UK).
9. *R. M. Malkani* v. *State of Maharashtra*: (1973) 1 SCC 471.
10. *Ziyauddin Burhanuddin Bukhari* v. *Brijmohan Ramdass Mehra & Ors.*: (1976) 2 SCC 17.
11. *State (NCT of Delhi)* v. *Navjot Sandhu*: 8 (2005) 11 SCC 600.
12. *Tukaram S. Dighole* v. *Manikrao Shivaji Kokate*:(2010) 4 SCC 329.
13. *Anvar P.V.* v. *P.K. Basheer & Ors.:* (2014) 10 SCC 473.
14. *Tomaso Bruno and Anr.* v. *State of Uttar Pradesh*: (2015) 7 SCC 178.
15. *Sonu alias Amar* v. *State of Haryana*: (2017) 8 SCC 570.
16. *Shafhi Mohammad* v. *State of Himachal Pradesh*: (2018) 2 SCC 801.

17. *Arjun Panditrao Khotkar* v. *Kailash Kushanrao Gorantyal*: (2020) 7 SCC 1.
18. *Ravinder Singh alias Kaku* v. *State of Punjab*: (2022) 7 SCC 581.
19. *State of Karnataka* v. *T. Naseer and Ors*: MANU/SC/1217/2023.
20. Federal Rules of Evidence, Retrieved from: https://www.law.cornell.edu/rules/fre.
21. *Zubulake* v. *UBS Warburg LLC*, 220 F.R.D. 212 (S.D.N.Y. 2003).
22. *United States* v. *Mitra-Hernandez*, 405 F.3d 492 (7th Cir. 2005).
23. *Lorraine* v. *Markel American Ins. Co.*, 241 F.R.D. 534 (D. Md. 2007).
24. *U.S.* v. *Vosburgh*, 602 F.3d 512 (3d Cir. 2010).
25. *United States* v. *Jones*, 565 U.S. 400 (2012).
26. *Riley* v. *California*, 573 U.S. 373 (2014).
27. Civil Evidence Act 1995, Retrieved from: https://www.legislation.gov.uk/ukpga/1995/38/contents.
28. Criminal Justice Act 2003, Retrieved from: https://www.legislation.gov.uk/ukpga/2003/44/contents.
29. Data Protection Act 2018, Retrieved from: https://www.legislation.gov.uk/ukpga/2018/12/contents/enacted.
30. The Civil Procedure Rules 1998, Retrieved From: https://www.legislation.gov.uk/uksi/1998/3132/contents.
31. The Criminal Procedure Rules 2020, Retrieved From: https://www.legislation.gov.uk/uksi/2020/759/contents/made.
32. Regulation of Investigatory Powers Act 2000, Retrieved From: https://www.legislation.gov.uk/ukpga/2000/23/contents.
33. *R* v *Bowden* [2005] EWCA Crim 2270.
34. *Barclays Bank Plc* v. *Devonport Royal Dockyard Ltd* [2008] EWHC 1509 (QB).
35. *Serco Ltd* v. *Secretary of State for Defence* [2011] EWHC 348 (TCC).
36. *Dawson-Damer & Ors* v. *Taylor Wessing LLP* [2017] EWCA Civ 74.
37. *Alibi* v. *CitySprint UK Ltd* [2019] ETCA 68.
38. Evidence Act 2006 No 69, Retrieved from: https://www.legislation.govt.nz/act/public/2006/0069/latest/DLM393463.html.
39. Electronic Transactions Act 2002 (2002 No 35), Retrieved from: https://www.legislation.govt.nz/act/public/2002/0035/latest/DLM154185.html.

40. Contract and Commercial Law Act 2017 (2017 No. 5) Retrieved from: https://www.legislation.govt.nz/act/public/2017/0005/latest/DLM6844033.html?search=sw_096be8ed81df44a8_electronic_25_se&p=1#DLM6844433.
41. Criminal Disclosure Act 2008 (2008 No 38), Retrieved from: https://www.legislation.govt.nz/act/public/2008/0038/latest/DLM1378803.html?search=sw_096be8ed81de6247_electronic_25_se&p=1.
42. Search and Surveillance Act 2012 (2012 No 24), Retrieved from: https://www.legislation.govt.nz/act/public/2012/0024/latest/DLM2136536.html.
43. Privacy Act 2020 (2020 No 31), Retrieved from: https://www.legislation.govt.nz/act/public/2020/0031/latest/LMS23223.html.
44. Evidence Act 1995 (No. 2, 1995), Retrieved from: https://www.legislation.gov.au/C2004A04858/latest/text.
45. Electronic Transactions Act 1999 (Act No. 162 of 199), Retrieved from: https://www.legislation.gov.au/C2004A00553/latest/text.
46. Privacy Act 1988 (No. 119, 1988), Retrieved from: https://www.legislation.gov.au/C2004A03712/2014-03-12/text.
47. The Notifiable Data Breaches (NDB) scheme, Retrieved from: https://www.oaic.gov.au/__data/assets/pdf_file/0011/5213/the-ndb-scheme-an-overview.pdf.

Index